ENVIRONMENTAL CRIMINOLOGY: SPATIAL ANALYSIS AND REGIONAL ISSUES

ADVANCES IN SUSTAINABILITY AND ENVIRONMENTAL JUSTICE

Previously ADVANCES IN ECOPOLITICS
Series Editor: Liam Leonard

PUBLISHED UNDER SERIES TITLE '**ADVANCES IN ECOPOLITICS**'

Transnational Migration, Gender and Rights Volume 10
Volume Editor: Ragnhild Sollund
Series Editor: Liam Leonard

PUBLISHED UNDER SERIES TITLE '**ADVANCES IN SUSTAINABILITY AND ENVIRONMENTAL JUSTICE**'

International Business, Sustainability and Corporate Social Responsibility Volume 11
Edited by Maria Alejandra Gonzalez-Perez and Liam Leonard

Principles and Strategies to Balance Ethical, Social and Environmental Concerns with Corporate Requirements Volume 12
Edited by Maria Alejandra Gonzalez-Perez and Liam Leonard

Environmental Philosophy: The Art of Life in a World of Limits Volume 13
Edited by Liam Leonard, John Barry, Marius De Geus, Peter Doran and Graham Parkes

The Sustainability of Restorative Justice Volume 14
Edited by Paula Kenny and Liam Leonard

Occupy the Earth: Global Environmental Movements Volume 15
Edited by Liam Leonard and Sya Buryn Kedzior

The UN Global Compact: Fair Competition and Environmental and Labour Justice in International Markets Volume 16
Edited by Maria Alejandra Gonzalez-Perez and Liam Leonard

Beyond the UN Global Compact: Institutions and Regulations Volume 17
Edited by Liam Leonard and Maria Alejandra Gonzalez-Perez

Lessons from the Great Recession: At the Crossroads of Sustainability and Recovery Volume 18
Edited by Constantin Gurdgiev, Liam Leonard and Maria Alejandra Gonzalez-Perez

Climate Change and the 2030 Corporate Agenda for Sustainable Development Volume 19
Edited by Maria Alejandra Gonzalez-Perez and Liam Leonard

ADVANCES IN SUSTAINABILITY AND ENVIRONMENTAL JUSTICE VOLUME 20

ENVIRONMENTAL CRIMINOLOGY: SPATIAL ANALYSIS AND REGIONAL ISSUES

BY

LIAM LEONARD
California State University, Fullerton, CA, USA

emerald PUBLISHING

United Kingdom – North America – Japan
India – Malaysia – China

Emerald Publishing Limited
Howard House, Wagon Lane, Bingley BD16 1WA, UK

First edition 2018

Reprints and permissions service
Contact: permissions@emeraldinsight.com

British Library Cataloguing in Publication Data
A catalogue record for this book is available from the British Library

ISBN: 978-1-78743-378-6 (Print)
ISBN: 978-1-78743-377-9 (Online)
ISBN: 978-1-78743-458-5 (Epub)

ISSN: 2051-5030 (Series)

Printed and bound by CPI Group (UK) Ltd, Croydon, CR0 4YY

ISOQAR certified
Management System,
awarded to Emerald
for adherence to
Environmental
standard
ISO 14001:2004.

ISOQAR
REGISTERED
Certificate Number 1985
ISO 14001

INVESTOR IN PEOPLE

CONTENTS

LIST OF CONTRIBUTORS

Hugo Aguas	California State University, Los Angeles, CA, USA
Dinur Blum	University of California, Riverside, CA, USA
Kevin Bucciero	California State University, Los Angeles, CA, USA
Christian Gonzalez Jaworski	University of California, Riverside, CA, USA
Anthony Keating	Edge Hill University, Faculty of Health and Social Care, Lancashire, UK
Liam Leonard	California State University, Fullerton, CA, USA
Shane Leonard	National University of Ireland, Galway, Ireland; Atlantic International Institute, Galway, Ireland
Darren McCauley	Arctic Research Centre, School of Geography and Sustainable Development at St. Andrews University, St. Andrews, Scotland
Brian McKenna	College of Arts, Sciences and Letters at the University of Michigan-Dearborn, Michigan, MI, USA

SERIES EDITOR'S INTRODUCTION

Having founded the Ecopolitics website in 2004, and then working with Emerald Publishing to produce both the *Advances in Ecopolitics* and *Advances in Sustainability and Environmental Justice* book series since 2007, I am pleased to present this, the 20th volume in this series. This collection focuses on research which combines a spatial analysis with environmental criminology, mapping crime, and deviance across different locations globally.

In the chapter "Framing Injustice in Green Criminology: Activism, Social Movements and Geography," Darren McCauley examines the manner in which a geographical perspective can assist with understandings of injustice. This spatial analysis looks at the social movements which emerge, and is framed in a green criminological context. The chapter "Anthropology at the Red-Green Crossroads" is located in an anthropological paradigm, as Brian McKenna opens up the debate between Green and Red ideologies. With progressive politics at a crossroads, these debates become all the more crucial for our society.

The chapter "The Ferguson Shooting, 2014: A Spatial and Media Analysis" is my own analysis of the shooting of African-American youth Michael Brown by police in Ferguson, St. Louis in 2014. This chapter incorporates both spatial and media analysis in order to open up the accounts of what happened in this tragic incident. This is followed by Dinur Blum's and Christian Gonzalez Jaworski's study of Spatial Patterns of Mass Shootings in the United States, between 2013 and 2014 in the chapter "Spatial Patterns of Mass Shootings in the United States, 2013–2014."

The chapter "Homeless Demography in Los Angeles County" sees Hugo Aguas present his Homeless Demography of Los Angeles County. Aguas contextualizes the issue of homelessness in Los Angeles, where those marginalized by homelessness are labeled as deviants by mainstream society. In the next chapter Anthony Keating looks at Police Culture, Gender and Crime in the Irish Free State in 1929. This historical account outlines the problems of deviant policing in the emerging Irish state, providing a regional analysis of sex crimes by those charged with upholding the law.

In the chapter "A Spatial Analysis of Crime: 'The Wire' and Depictions of Urban Crime," Shane Leonard provides an analysis of urban criminology through an analysis of the cult television series "*The Wire*." This spatial analysis is located in Baltimore, where local ethnic communities are outlined within the frameworks of the overlapping challenges which they face. In the book's final chapter, the issue of illicit drug use is examined by Kevin Bucciero's spatial analysis of methamphetamine use in North America. Ultimately, I am pleased to see such a collection of research from both experienced and early career academics. This bodes well for the future of research in this field, and I look forward to many more volumes in the Advances in Sustainability and Environmental Justice Series in the coming years.

Liam Leonard
Series Editor

FRAMING INJUSTICE IN GREEN CRIMINOLOGY: ACTIVISM, SOCIAL MOVEMENTS AND GEOGRAPHY

Darren McCauley

ABSTRACT

Injustice is perceived, experienced and articulated. Social movements, and their constitutive parts, frame and re-frame these senses of injustice. Two often-overlapping accounts of social movements are in focus in this chapter. Human geography has been flooded with movement-based analyses of environmental justice (EJ). Sociology (more appropriately political sociology) has provided insight into social movements in the form of 'contentious politics' (CP). Building on both sets of literature, this chapter seeks to advance thought in human geography through a detailed exploration of master and collective action framing. It argues, firstly, that framing analysis challenges activist researchers to retain 'spatial constructs' as their central focus, rather than discourse. It calls, secondly, for us to unbind injustice *as much as justice in our analysis of framing. And lastly, it demands a multi-spatial perspective on framing beyond simply scalar accounts.*

Keywords: Social movements; environmental justice; contentious politics; framing; scale; injustice

Environmental Criminology: Spatial Analysis and Regional Issues
Advances in Sustainability and Environmental Justice, Volume 20, 1–20

ISSN: 2051-5030/doi:10.1108/S2051-503020170000020001

GEOGRAPHY, SOCIAL MOVEMENTS AND FRAMING

> We believe it makes a difference not just where our studies and stories go, but also from whence they come. (Reed & George, 2011, p. 840)

The notion of framing has emerged in geographical thought set within environmental justice (EJ) scholarship (Harrison, 2006; Heynen, 2003; Kurtz, 2003; Lieshout, Dewulf, Aarts, & Termeer, 2011; Sze et al., 2009; Towers, 2000) from a well-established body of political sociology or 'contentious politics' (CP) literature (Baud, 2005; Benford & Snow, 2000; Boykoff & Laschever, 2011; Della Porta & Mosca, 2007; Diani, 2000; Gerrards & Rucht, 1992; Goffman, 1974; Leitner, Sheppard, & Sziarto, 2008; McGammon, Muse, Newman, & Terrel, 2007; Powell, 2011; Zald, 2000). Both sets of authors set out to understand, from a notably social psychological perspective, the role of social movements. In so doing, they have developed, expanded and tested the collective framing of discourses. The development of 'scale frames' has allowed us to more effectively deal with a complicated set of multi-scalar relationships of contradictions and dependencies (Kurtz, 2003). Framing challenges, moreover, geographers to 'foreground spatial constructs – that is human and non-human agents – rather than discourses' (Bickerstaff & Agyeman, 2009, p. 783). This chapter seeks, in part, to enable an inter-disciplinary social movement response to the call for 'imaginative' EJ research (Holifield, Porter, & Walker, 2009, p. 601). It brings together state-of-the-art research and thought on EJ and social movements through a deeper elaboration of framing processes and a focus on injustice, rather than justice.

At the heart of this chapter, a critical disagreement must be placed upfront. For practice-based insight to flourish we should not always accept the dominant understanding of the idea of praxis – that is that theory and practice must inform each other (Sze & London, 2008; Sze et al., 2009). In its place, we should be willing to adopt an understanding whereby movement experience informs theory, but not necessarily the other way around. In such a conceptualization, theory does have a role in exploring normative-based consequences and implications. It should not, however, drive our assessments of movement experiences. This chapter does not quite go as far as Debanne and Keil in arguing that 'movements linked to justice struggles are highly diverse and cannot be measured or expressed in universal terms' (2004, p. 209). In line with Schlosberg's comments, '(t)he point is that different discourses of justice, and the various experiences and articulations of injustice, inform how the concept is used, understood, articulated and demanded in practice; the engagement with what is articulated on the ground is of crucial value to our understanding and development of the concepts we study' (2013, p. 50).

There are two observable trends in social movement literature in geography that this chapter seeks to reflect upon. The theorization and application of justice concepts has moved our attention away from 'activism' towards normative-based 'analysis'. To build a little on this rather simplistic distinction, there has been a shift from assessing and understanding the viewpoints of activists (and as argued below 'non-activists') towards the analysis of regimes as normatively just (Holifield, 2004; Walker & Day, 2012) due to, what Lievianos refers to as, the increasing 'state resonance of environmental justice'[1] (2012, p. 491). This 'analytical' (or rather normative) turn threatens to severely limit research in social movements to confirming or disproving the application of theoretically developed and selected frames. It is argued here that a focus on understanding injustice, rather than applying normative models of justice, offers a way out of this straightjacket. The often-coined phrase comes to mind: 'whose justice?' – or more appropriately here 'injustice in whose eyes?' Framing is (only) one attempt to bring the conceptualization of justice back to the protagonists of environmental activism.

The second trend in this literature, in light of the 'multiple spatialities' justice research (Walker, 2009a, 2009b, 2012), is the move from vertical to horizontal accounts of space. This has, firstly, amounted to critiques on the usefulness of scale in human geography, most notably from Marston, Jones, and Woodward (2005). They comment, 'we are convinced that the local-to-global conceptual architecture intrinsic to hierarchical scale carries with it presuppositions that can delimit entry points into politics' (2005, p. 427). Inspired by CP scholarship, it is argued below that scale is rarely the entry point for social movement scholars. In reality, individuals, organizations, states and even non-human agents or more appropriately 'spatial constructs' (Bickerstaff & Agyeman, 2009) represent the starting point in our understanding. In so doing, social movement action is no longer confined to scalar entry points. In support of Moore (2008), scale in this context is a category of practice rather than an analytical category in itself. As explored below, the real concern with the politics of scale is not the starting point, but rather the end. This turn to 'all things horizontal' is, therefore, equally refuted as an attempt to 'throw the baby out' of a warm bath into a rather choppy sea. In other words, research on social movements must retain (and develop further) scale frames – but as only one type of frame.

The concept of framing as used in CP scholarship is derived from the work of Goffman (1974).[2] A frame denotes 'schemata of interpretation' that allow individuals to 'locate, perceive, identify and label' occurrences within their 'life space' (Goffman, 1974, p. 21). This entails agency in the sense that what is evolving is the work of social movement activists or movement activists. 'Master' frames designate, firstly, the overall signifier for mobilization in and across movements. They are often adopted early on in the cycle of protest and, as a result, set up a durable inclusive and excluding framework for social mobilization (Snow & Benford, 1992). It is argued below that master frames of 'injustice' should replace the normative and theoretically laden frame of 'environmental justice'. Collective action frames are, secondly, 'action-oriented' sets of beliefs and meanings that are designed by activists to inspire and legitimize types of action. The geographical literature in this area remains constrained by the construction of 'scale frames'. This chapter suggests an equal need to explore 'place' and 'network' frames.

We must, therefore, place the agents of such claims at the forefront of social movement research. Geographical scholarship on EJ has tended to prioritize the discourse and the meanings it produces (Agyeman, 2002; Davies, 2006). In contrast, CP research on activism prioritizes the individuals and organizations involved (Boykoff & Laschever, 2011; Powell, 2011). From this perspective, it considers movements or movement actors (organizations and activists) as signifying agents engaged in the production and maintenance of meaning for protagonists, antagonists and bystanders. Framing demands, therefore, the foregrounding of spatial constructs in framing practices, rather than the discourses themselves. The continuous reformation, contestation and re-production of frames remind us that they are not static reified entities. They are, in fact, deployed to legitimate movement goals and campaigns (Benford & Snow, 2000; Heynen, 2003; Kurtz, 2003; Sze et al., 2009; Towers, 2000). Framing allows us to question more effectively the key agents, or more precisely the active role of spatial constructs in framing practices.

Social movements by their very nature are entities that aim to remedy a perceived injustice. Benford and Snow (2000) refer to three key components that are always present in framing processes. Attributional framing accords blame and responsibility, while prognostic framing involves the articulation of a proposed solution. Motivational framing refers, thirdly, to the construction of appropriate vocabularies of motive (Benford, 1993). Gamson (1992) and Gamson, Fireman, and Rytina (1982) consider that injustice is always at the centre of these framing processes – a conclusion tempered by Benford and Snow (2000). Beckwith comments, 'collective action frames, in organizing and making sense of lived experience and perceptions, serve to identify social injustices, to focus and to summarize grievances, to organize disruptive action and to express disruption, and to posit opposition and solutions' (2001, p. 301).

Environmental problems are, after all, socially constructed claims defined through collective processes (O'Brien, 2011; Taylor, 2000). Di Chiro comments, '(d)efining what counts as an environmental problem and what doesn't ... delegates different issues as either inside or outside the environmental "frame"' (2008, p. 279). Claims for justice are often broad and diverse (Fan, 2006). Framing analysis needs to reflect these multiple understandings, rather than any pre-set theories or assumptions. The vast majority of EJ literature concentrates on (and often limited to) 'master' frames for collective action such as 'just sustainability' (Agyeman, 2005) or the 'environmental justice paradigm' (Taylor, 2000). A master frame can expound a series of normative and practical solutions. It is, nonetheless, at its very heart, the mobilizing 'call to arms' for communities in the face of a perceived injustice. The fluidity of master frames of injustice offers a way out from justice-based conceptualizations.

*IN*JUSTICE 'UNBOUND'

Injustice – rather than justice – should be the focal point for EJ research through a more explicit assessment of master frames of 'injustice'. Master frames are collective action frames that have expanded in scope and influence. Put simply, a master frame encompasses the contextual boundaries, interaction and normative claims of more than one organization or one movement. Such frames can indeed vary dramatically in terms of restrictiveness or exclusion. Gerhard and Rucht (1992) found that two distinct master frames (with different protagonists, antagonists, organizations, etc.) worked together to encourage social mobilization in Germany. They can, therefore, often serve as a 'kind of master algorithm that colors and constrains the orientations and activities of other movements' (Snow, 2004). It is argued below that geographical scholarship in EJ research remains theoretically, conceptually and contextually bound. This section concludes with a reflection on not only unbinding EJ research from pre-set notions of justice, but also its conceptualization of 'environment'.

Theoretical accounts of EJ threaten, *firstly*, to bind social movement researchers into pre-determined logics of justice (Barnett, 2010). For Caney (2010), justice-based activism research has hitherto focused on exposing and proposing archetypal normative frameworks. In support of Agyeman, Cole, Haluza-DeLay, and O'Riley (2010), Reed and George comment, 'researchers are cautioned that the long-observed disconnect between theory and practice in the field of environmental justice may be exacerbated should academics become more concerned with theoretical refinement over progressive, practical, and possible change' (2011, p. 839). The theorization of justice seeks to expose ideal end points (and more recently processes) from various philosophical traditions. For example, Okereke (2006) finds that any notions or principles of justice originate from five distinct incarnations: utilitarianism, communitarianism, liberal equality, justice as meeting needs and libertarianism – later refined to include 'market justice' (Okereke & Dooley, 2010). In a similar vein, Schlosberg (2004, 2013) argues that justice theorists need to be pluralist in accepting a range of understandings of 'good'. It is argued here that we need to instead explore the plurality of injustice.

The first step in this direction is indeed the acknowledgement that the study of justice is pluralist. Martin, Gross-Camp, Kebede, McGuire, and Munyarukaza acknowledge, 'that justice poses considerable conceptual challenges, not least because of the practical (if not intellectual) impossibility of reaching consensus' (2013, p. 2). This is borne out by a valiant theoretical sortie through the myriad of approaches to conclude that justice is both plural and multi-dimensional. Their conclusion bears a self-reflective unease; 'we clearly have much to learn about the limitations of our own framing and methods, including our inevitable starting point in logics of justice' (2013, p. 10). The second move involves an acknowledgement that justice is contextualist whereby some principles may apply in certain situations. Walker comments, 'as we move from concern to concern *and* from context to context, we can expect shifts in both the spatial relations that are seen to be significant and in the nature of justice claims being made' (2009a, p. 622).

Ideal justice theorists seek to effectively eliminate the potential for conflict. Schlosberg comments however, 'such theorists are mistaken ... (c)onflicts of justice arise ... problem solving entails the negotiation of different conceptions of (in)justice in and across participants, from community or stakeholder groups to corporation or states' (2013, p. 45). Schlosberg claims that the idea of environmental justice has 'examined multiple reasons for the construction of *injustice*' (2013, p. 37). This chapter calls, however, for an exploration of the construction of multiple injustices. The expansion in the theorization of environmental justice as a concept must be answered with a similar response in our understanding of environmental activism. As Barnett comments in support of Sen (2009),

> Rather than thinking of philosophy as a place to visit in order to find idealised models of justice or radically new ontologies, we would do well to notice that there is an identifiable shift among moral and political philosophers towards starting from more worldly, intuitive understandings of injustice, indignation, and harm, and building up from there. (2010, p. 252)

The recent development of normative concepts of justice looms, *secondly*, in a similar manner. There is a sense (to some extent correctly) that such concepts are worldly, emerging from situated conflict. They are, however, more often emergent from philosophical debate. A set of normative-based testable assumptions materialize based upon achieving equity and fairness in the distributional, post-distributional – referred to as 'recognition' largely attributed to Nancy Fraser (1997)[3] and developed by Schlosberg (2004, 2007, 2013) – and procedural burdens of environmental risk. Distributional justice suggests that people of colour and/or low income should not bear a disproportionate burden of pollution (Harvey, 1996). On 'post-distributional' justice, Gibson-Wood and Wakefield comment, '(a) lack of recognition is unjust because it denies an equal voice or "place at the table" to those whose understanding or experiences of "the environment" or "environmentalism" is outside of dominant understandings or experiences' (2013, p. 645). McCauley, Heffron, Stephan, and Jenkins (2013) refer to this framework as the 'triumvirate of justice tenets' (distributional, procedural and recognition).

Gibson-Wood and Wakefield (2013) employ this triumvirate as frames, and demonstrate how it can act as one insightful 'master' frame for understanding environmentalism among Hispanics in Toronto. In other words, each tenet provides a set of expectations for what should be achieved against what actually happened. The 'weakest link' is then identified (in this case recognition). The analytical objective identification of injustice can be blind to the experiential perception of spatial constructs. The more recent attempt to uncover a third form of EJ tenets as the 'post-distributive justice of recognition' (Bulkeley,

Carmin, Castan Broto, Edwards, & Fuller, 2013) threatens, for example, to unintentionally disrobe those who are unrecognized of any meaningful agency. Even though Fraser (2008) firmly identifies social movements as key agents of change, the emphasis is on the call for 'authorities' and 'policy-makers' to recognize under-represented groups – such as in Walker and Day (2012). Framing research emphasizes, in contrast, the need to explore such processes among those who are 'under-recognized' in order to gain insight into the success or not in mobilizing injustices. They are referred to below, not as 'victims', but rather 'non-activists' as a new challenge for social movement and justice research.

Our approach to EJ remains, *thirdly*, contextually bound. In this vein, the EJ 'master' frame is derived from specific empirical contexts. The origins of EJ research are accepted to be race and poverty-based campaigns involving multiple organizations and individuals across the US merging into a veritable EJ movement – often cited as beginning in Warren County, North Carolina (Bullard, 1999; Szasz, 1994; Urkidi & Walter, 2011). And thus, the EJ master frame in the United States is formed around race, class, gender and the environment. Taylor (2000) talks explicitly about the 'environmental justice paradigm' as a master frame which links together 'environment, race, class, gender and social justice' issues. In the United Kingdom (especially among NGOs), the master frame has been termed as 'just sustainability' (Agyeman, 2005; Agyeman, Bullard, & Evans, 2003; Agyeman & Evans, 2004) despite the earlier observation that there exist 'at least three different constructions of environmental justice' (Agyeman, 2002, p. 37). This refers to a frame that links together issues of sustainability, social inclusion and procedural equity (Bikerstaff & Agyeman, 2009, p. 782).

Dawson (2000) demonstrates, however, the potential fluidity of EJ master frames in linking it explicitly to eco-nationalism. She identifies sub-group identity, social justice and environmentalism as the core tenets in the US EJ frame. The US environmental movement is, in her view, built on the foundation of sub-group identity and the desire for social justice. As a result, groups defined by religion, gender, national identity or class could offer a basis for EJ movements and their master frame. In this way, the EJ frame covers, for example, the protection of indigenous peoples across the Americas (Cantzler, 2007; Holifield, 2012; Schlosberg & Carruthers, 2010; Urkidi & Walter, 2011) or Taiwan (Chi, 2001) or tribal groups from environmental hazards in Africa (McDonald, 2002; Visser, 2003). In such a conception, the EJ frame can actually be ultimately divisive and exacerbate violent conflict. Dawson traces the environmentalist roots of nationalist movements in the former USSR leading directly to social tensions and fragmentation. She observes, 'the intertwining of environmental causes and sub-group identities can be seen to both enhance environmental mobilisation among previously immobilized groups and deepen a pre-existing sentiment of "us" versus "them" within the population' (2000, p. 36).

Empirical-based conceptions of justice are, therefore, as problematic as theoretical and conceptual incarnations. Pellow and Brulle argue, indeed, that '(s) cholars cannot understand ... environmental injustices through a singularly focused framework that emphasizes one form of inequality to the exclusion of others' (Pellow & Brulle, 2005, p. 298). Our attention should be drawn to where and when injustice is felt and experienced. Hobson (2006) argues indeed that EJ research must diversify its understanding of where injustice can be found. In her assessment of an environmental organization in Singapore, she demonstrates how environmental injustice is felt in everyday practices of individuals and organizations, even where expressions of public concern on the environment are infrequent or at least highly managed. More recently, substantial research has focused our attention on injustices within climate activism (Barrett, 2012; Bulkeley et al., 2013). The fluidity of master frames on EJ offers one potential solution to unbinding how we approach justice and injustice. We now turn our attention to unlocking further how we can explore master frames of injustice through a better understanding of collective action framing.

MOVING *WITH* SCALE FRAMING

Framing is, above all, a strategic practice (Kurtz, 2003; Leitner, 2003; Leitner et al., 2008; Sze et al., 2009). In other words, collective action frames are deliberative, utilitarian and goal directed. Collective action frames originate from the notion of 'repertoires of contention'. In his book, *The Contentious French*, Charles Tilly refers to such repertoires as a 'set of routines that are learned, shared, and acted out through a relatively deliberate process of choice ... where people know the general rules of performance more or less well and vary the performance to meet the purpose at hand' (1986, pp. 390, 392). These repertoires are developed through a process of learning and practice, importation of experiences from other social movements and the mobilization of new constituencies. Change is achieved, therefore, through deliberate innovation and strenuous bargaining where 'people create, adapt, apply and deploy shared understandings' (Tilly, 1995, p. 44). Changes in collective action repertoires, therefore, depend upon the frames of meaning used by strategic actors. In other words, movement activists undertake active signifying work that implies agency and contention at the level of reality construction (Snow & Benford, 1992).

Collective action frames are essentially the properties of organizations or groups of individuals (Benford & Snow, 2000).[4] In short, NGOs, interest groups, pressure groups, environmental groups, etc., 'frame, or assign meaning to and interpret relevant events and conditions in ways that are intended to mobilize potential adherents and constituents, to garner bystander support, and to demobilize antagonists' (Snow & Benford, 1988, p. 198). Firstly, they set the contextual boundaries in so far as what is effectively 'in' and 'outside' the collective action frame in terms of perceived antagonists, events, core issues and a timeline. This frame is, secondly, articulated by a restricted set of individuals set within and alongside the organization in question (Baud, 2005; Goffman, 1974; Powell, 2011). Collective action frames involve, thirdly, normative claims on how problems should be approached. The Tea Party movement in the United States involved a wide range of organizations prioritizing different collective action frames that made central normative claims on abortion, the economy, appealing to the 'everyday American', avoiding 'flash in the pan', etc. (Boykoff & Laschever, 2011).

This often leads to competing frames between organizations, and more often contentiously challenging existing authoritative views and framings of reality (Cantzler, 2007; Diani, 1996). In this way, social movement actors are involved in a politics of signification internally between activists as well as with opponents (referred to by Kurtz, 2003 and Cantzler, 2007 as 'counter-frames'). Within this context of framing processes, CP researchers consider political opportunity structures as determinants of social action (Giugni, 2011). In a sea of definitional contestation, we summon Koopmans' tellingly simplistic designation as 'factors outside the mobilizing groups' (2004, p. 63). As Meyer comments, 'the key recognition in the political opportunity perspective is that activists' prospects for advancing particular claims, mobilizing supporters and affecting influence are context-dependent' (2004, p. 126). Both CP and EJ scholars have, in this regard, sought to expunge the primacy of the state[5] in social movement research in revealing multi-scalar, place-specific and networked (Tarrow & della Porta, 2005) action that transcend the state apparatus.

The origin of the scalar perspective as one response to state-centric opportunity structures indicates, for CP scholars, a lasting durability in this research area (Tarrow & McAdam, 2004). For EJ research, Williams (1999) identifies a first wave of EJ literature that focuses on reified representations of scale. A second wave of scholarship is characterized by a more pronounced ability to accommodate the social construction of scale (Bickerstaff & Agyeman, 2009; Kurtz, 2003; Paasi, 2004). Many scholars have demonstrated thus that actors employ scales in various beneficial ways within the context of multiple interactions taking place at different scales (Delaney & Leitner, 1997; Harrison, 2006; Kurtz, 2003; Lieshout et al., 2011; Marston, 2000; Sze et al., 2009; Towers, 2000). Leitner (1997) demonstrates how nationalist right-wing parties manage to employ different scale frames in opposition to the concentration of immigration policy at the supra-national scale. Such research finds that movements attempt to maximize the local scale whilst tapping into the supranational or global. Harvey reminds us further, 'the choice of spatial scale is not "either/or" but "both/and" even though the latter means confronting serious contradictions' (2000, p. 51).

Kurtz (2003), Harrison (2006) and Sze et al. (2009) use the notion of scale frames to encapsulate the discursive practices that construct links between the scale at which a social problem is experienced and the scale(s) at which it could be solved. Williams reminds us that '(a) dynamic of scale politics centres on an antagonistic relationship between a societal problem and its political resolution ... (t)he scale at which a social problem is generated may not coincide with the scale(s) at which the problem might be resolved' (1999, p. 56). In so doing, such authors recall the attributional (i.e. blame and problem construction) and prognostic (solution) framing processes outlined above (Benford, 1993; Gamson, 1992; Gamson et al., 1982). They fail, however, to substantially comment on the third process of motivational[6] (i.e. identifying vocabularies of motive) framing, with the notable exception of Martin (2003) with regards to 'place-framing' (explored below).

Advances in this area of geographical research have sought to escape the upscaling argument. Kurtz (2003) shed light on the role of counter-scale frames that seek to constrain upscaling. In a similar fashion, Harrison (2006) demonstrates, for example, how regulators succeeded in 'downscaling' the framing of pesticide drifts as isolated accidents. Haarstad and Floysand (2007) demonstrate that constructing legitimate discourses on several scales plays a key role in successfully opposing mining in Peru. More notably, scholars have moved beyond scales of regulation to theorize scales of meaning (Towers, 2000) or inclusion/exclusion (Sze et al., 2009). In their exploration of agricultural practices in the Netherlands, Lieshout et al. (2011) break the notion of scale down into four distinct categories: spatial, administrative, agricultural and time. Beyond the expansion of scale, there remain important questions to be explored on transcalar issues from climate change to the genetic scale, as suggested by Bickerstaff and Agyeman (2009), or on hitherto under-explored 'motivational' framing process as revealed here.

MOVING *BEYOND* SCALE FRAMING

The second trend in social movement research is the prioritization of horizontal or flat understandings of space. Scale (albeit one 'vision' of space – as explored above) has been central in geographical approaches to movements. This section does not aim to reflect on the rights or wrongs of the politics of scale. Scale has been, and will remain, a pertinent frame for social movement activists. From this perspective, we refute the suggestion made by Marston et al. (2005) to replace scale with a flat ontology. For frame, and more broadly social movement analysts, scale does and will continue to matter. However, O'Brien warns, 'a narrow focus on scale can also lead to misinterpretations of cause and effect' (2011, p. 545). Above all, this chapter refutes the suggestion by Kurtz that 'the very concept of environmental injustice precipitates a politics of scale' (2003, p. 891). We should accept that, in some instances, the politics of scale may be replaced by alternative spatialities. Put differently, does a movement have to jump scales to succeed? In following, scale, and counter-scale, frames only represent one type of frame.

A 'third' wave of EJ scholarship based upon rejecting the primacy of vertical conceptions of space is observable on the back of a now increased ability to accommodate multiplicity, change and the social construction of scale. The move away from scales of regulation has allowed researchers to accept that scale frames are constructed across horizontal space. In support of Leitner et al. (2008), this observation underlines, firstly, the enduring but under-valued nature of place in framing. Contemporary scale theorists share with those who focus on place a common emphasis on the territorial nature of societal organization. At first sight, places that are separated at one scale become connected through their common association with a higher scale (Sheppard, 2002). However, place theorists deny the claim that scales connect territorially bounded entities. For them, place is in fact open and heterogeneously constituted (Massey, 2005; Verstraete & Cresswell, 2002). With this in mind, we need to respect 'place-framing' as a separate form of collective action framing from 'scale-framing'.

Places have a distinct materiality that mediates and regulates social relations and daily routines. This materiality of space forms the nature and possibility of contention and social action. Beckwith demonstrates (2001) how place constrained attempts of the United Mine Workers of America to achieve a shift in frame from traditionally violent to nonviolent forms of protest. The historical development of the mines in question was deeply embedded in place-specific constructions of masculinity and violence. Places are embued with meaning as well as power. In this vein, social movements attempt to strategically manipulate, subvert and resignify places in beneficial ways. Franquemanque (2007) and McCauley (2011, 2013) demonstrate how French anti-GMO activists reframed a proposed military base in 1971, Larzac, as both a physical site and emblem for a wide range of anti-globalization protests ever since. Heaney and Rojas (2006) explore how different organizations in the anti-Iraq War movement in North Carolina fought over the multiple meanings of place in competing strategic efforts to frame social movement activity.

Place can provide an important mobilizing discourse and identity for collective action. For place-framing, we should, therefore, examine *how* place appears in the discourses of organizations, and *why*. Martin (2003) demonstrates how movement organizations in Minnesota repeatedly cited 'neighbourhood' as a sphere of action with problems that residents could identify and understand. A similarly 'meso-level' approach (Reid, Sutton, & Hunter, 2010) could identify new research questions around not only neighbourhood as a signifying place for activists, but also the household (Barr & Gilg, 2006; Cress & Snow, 2000). Such an approach could, alternatively, be easily applied to recent advances in climate justice activism (Bulkeley et al., 2013) to investigate how activists frame and re-frame the 'city' as a place full of meaning and power. Such research does, indeed, remind social movement scholars that place matters as much as space.

So what of multiple spatialities? After all, for Gieryn, 'place is not space' (2000, p. 465). Leitner et al. (2008) suggests that the politics of mobility offers much in this regard. Tim Creswell comments, '(i)f movement is the dynamic equivalent of location, then mobility is the dynamic equivalent of place' (2006, p. 3). The new mobilities paradigm seeks to capture the meaning, power, practice and embodiment in the displacement of people. Appearing in unexpected places, social movements use mass demonstrations and rallies as well as bike and bus rides to their tactical advantage (Caren, Ghoshal, & Ribas, 2011). This experience of mobility shapes, moreover, the identities of activists. A less nascent conceptual sortie into explicitly horizontal spatialities is found in the politics of networking. Leitner and Miller consider networking as the most effective way for 'advocates of scale framing to deny the simplistic hierarchical power assumptions in dealing with scales' (2007, p. 119). In exploring the human rights group *Madres de Plaza de Mayo* in Argentina, Bosco (2001) observes that the network approach reveals how social relations are embedded in webs of meaning and practice in a way that better emphasizes the spatial relationship between place and space for social movement scholars.

The concept of networking in social movement research is positioned as a challenge to states (particularly for CP scholars). Sewell (2001), for example, points to the construction of activists' face-to-face and virtual communication networks as a means to navigating away from state and business-controlled incarnations. Trans-local networks have emerged to prevent movements from being limited spatially (Routledge, 2003). Wang and Soule (2012) demonstrate through a longitudinal study of protest activities that networking among movement organizations has resulted in broader tactical repertoires and increased repertoire usage through a process of 'tactical diffusion'. Della Porta and Mosca (2007) reveal in their study of the Italian global justice movement how networking develops our understanding of framing through emphasizing 'bridging' processes. In following, networking is considered as an 'antidote to single issue claims ... where participation in protest campaigns is reflected in the bridging of several issues and frames' (2007, p. 19). Inspired by the works of Bruno Latour, a recent expansion in the actor network theory (ANT) literature has sought to include non-human agency in justice research (Bickerstaff & Agyeman, 2009; Holifield, 2009, 2012).

Framing allows us, from this perspective, to question more effectively the key agents in network-centred explanations of the spatiality of scale for example spaces of dependence and engagement (Cox, 1998) or assemblages (Bickerstaff & Agyeman, 2009). After all, 'some sort of organization is at the centre of attempts to ... pursue the goal of constructing a network of associations' (Cox, 1998, p. 15). Leitner et al. support such work in calling for 'attention to how agency is distributed across the more-than-human world, and not solely located with humans' (2008, p. 158). Schlosberg (2013), especially through the capabilities approach (Schlosberg & Carruthers, 2010), considers the horizontal relationship between humans and non-humans as critical for EJ scholarship. For him, it effectively represents how we perceive human–environment interaction. It reminds us that environmental injustices are as much in the treatment of the non-human realm as in relations between human beings.

Our overall exploration of injustice must, therefore, seek to delve deeper into how movements frame the non-human world. Leitner et al. (2008) raises the issue of socio-positionality in an attempt to encourage social movement researchers to consider the role of non-human agency. The continual re-production of positionality should be considered alongside the role of non-human agents. Technology has, from this perspective, the ability to empower certain actors whilst disempowering others. Sheppard (2002) evokes the use of the term 'socio-spatial positionality' as a response to the downplay of power hierarchies in networking. Positionality often refers to the social situatedness of subjects in terms of gender, race, class, sexuality and other axes of social difference. At the same time, the co-constitutive nature of the social and space are implicated in the production and re-production of identity. As a result, a social movement's agenda is shaped by its socio-spatial positionality – the type of participants, where or who participates in various ways, their social values or even the geographical situatedness of the organization itself.

CONCLUSION

We have examined above the two dominant conceptualizations – master and collective action – of framing processes in social movement literature from both an EJ and CP perspective. This chapter suggests that we should turn our attention in master framing towards exploring injustice, rather than pre-set theoretical, conceptual or even contextual understandings of justice. The concept of master framing provides, in this way, an opportunity to explore the geographies of injustice. Collective action framing must, secondly, accommodate a wider agenda than scale framing in geographical thought, encompassing advances in place and network framing. In so doing, it needs to embrace new agendas such as mobilities, socio-spatial positioning and non-human agency. The overall approach to the notion of praxis (outlined above) is, indeed, rather simplistic. One could argue that framing is, itself, a pre-set theory. It does, however, challenge us with an important question: how can we listen better?

Houston (2013) challenges human geographers to explore the imaginative practices of how stories are performed in environmental justice struggles in shaping alternative imaginations of place. Her perception of injustice is firmly rooted in sites of pollution. For frame analysis, storytelling can be a powerful tool. It produces different environmental imaginaries about the kinds of worlds they want (or not) to live in. In so doing, such stories can act as a sort of mobile repertoire for combating environmental injustices in other places. They can effectively inspire the construction of frames based on success elsewhere. In referring to Aboriginal storytelling, Reed and George (2011) remind us, moreover, that readers needed to adopt new ways of listening in order to understand not only senses of justice but also injustice. This chapter challenges academics to similarly adopt, or as in several instances continue, an approach predicated on 'listening' to how injustices are experienced in relation to environmental concerns (for EJ) and the state in its various forms (for CP researchers). Schlosberg comments;

> Environmental justice movements have been challenging the discourse of development in the streets, in the media, and in the halls of the institutions of the global economy. But they also challenge our own discourse of justice in academia as well, and we would do well to listen in. (2004, p. 537)

We should not assume the origins of injustice. Both master and collective action framing illuminate the roots, ideas, values, identities and actions of organizations and various groupings of individuals. They allow us to investigate how the 'owners' of such frames perceive, experience and imagine the world. With this in mind, a framing agenda offers significant potential to investigate two emerging sets of agents in the EJ literature inspired by recognition justice and assemblage – 'non-activists'[7] and 'non-humans'. How do social movement researchers deal with those who do not speak or act? This chapter suggests that the bedrock of collective action frame may be misplaced – such a frame could belong to a set of performances/practices and assemblages as much as an organization. As demonstrated by Houston and Pullido (2002), performative justice allows 'insights into ... individuals and groups not able or willing to engage in overt environmental struggles' (Hobson, 2006, p. 674). Bickerstaff and Agyeman (2009) reveal, moreover, the potential of a dual framing and assemblage approach. How performance and assemblages frame such individuals' and groups' perceptions of injustice is a matter for future framing studies.

The multiple spatialities of justice represents more than the now well-trodden observation that 'we need to move beyond the Cartesian view of distributive justice'. This chapter alludes to the multiple spatial framing of social movements. Such framing processes must continue to exude the key attributive, prognostic and motivational traits as found in collective action framing literature. Scale-based framing is only one albeit important account of movement behaviour in this area, arousing new spatial accounts whilst respecting the continuing and overlooked validity of place-framing. In order to move debates forward, EJ and CP scholarship must equally advance how geographical thought deals with the notion of political opportunity structures. EJ research maintains an awkward relationship with CP scholars such as Diani (2005) and Tarrow (2005) who firmly believe social movements remain embedded in national and state-centric contexts. In doing so, this chapter calls for future work on 'multiple spatial opportunities', grappling with the spatialities of Walker and Schlosberg, the political opportunities imagination of Giugni and Meyer as well as emerging innovations such as discursive opportunity structures (McGammon et al., 2007). Barnett comments,

> The emphasis on the situations that generate political action requires us to supplement the focus upon processes of spatial extension that underwrites the work of Fraser, Dryzek, and Bohman, with a consideration of contexts which generate the modes of transnational agency they focus upon. (2012, p. 684)

NOTES

1. Lievianos (2012) claims that EJ movements' concerns have become captured and institutionalized by the state apparatus due to their increasing 'resonance'. In following, EJ-based research has been attracted to examining how appropriately (or not) the various guises of the state are dealing with and implementing conceptualizations of EJ.
2. Please see Shmueli (2008) for a much wider discussion on framing processes beyond social movement literature.
3. Please see also Fraser and Honneth (2003).
4. This statement is challenged in the conclusion with reference to performative justice.
5. Williams and Mawdsley (2006) acknowledge, however, that EJ studies in geography must take better account of the state's role (albeit with reference to specifically developing states), even if its influence has diminished somewhat in explanations of social mobilization.
6. Please see Wright and Boudet (2012) for a detailed exploration of motivation and social movements.
7. EJ literature on recognition has tended to see a victimised inactive community as one or several homogeneous and united group(s) of people. In support of Ishiyama (2003), the term 'non-activists' challenges us to define more clearly agency in 'under-recognised' groups.

REFERENCES

Agyeman, J. (2002). Constructing environmental (in)justice: Transatlantic tales. *Environmental Politics*, *11*(3), 31–53.

Agyeman, J. (2005). *Sustainable communities and the challenge of environmental justice*. New York, NY: New York University Press.

Agyeman, J., Bullard, R., & Evans, R. (2003). *Just sustainabilities: Development in an unequal world*. Cambridge, MA: MIT Press.

Agyeman, J., Cole, P., Haluza-DeLay, R., & O'Riley, P. (2010). *Speaking for ourselves: Environmental justice in Canada*. Vancouver: UBC Press.

Agyeman, J., & Evans, B. (2004). Just sustainability: The emerging discourse of environmental justice in Britain? *Geographical Journal*, *2*, 155–164.

Barnett, C. (2010). Geography and ethics: Justice unbound. *Progress in Human Geography*, *35*(2), 246–255.

Barnett, C. (2012). Situating the geographies of injustice in democratic theory. *Geoforum*, *43*, 677–686.

Barr, S., & Gilg, A. (2006). Sustainable lifestyles: Framing environmental action in and around the home. *Geoforum*, *37*, 906–920.

Barrett, S. (2012). The necessity of a multiscalar analysis of climate justice. *Progress in Human Geography*, *37*(2), 215–233.
Baud, M. (2005). *Popular intellectuals and social movements: Framing protest in Asia, Africa and Latin America*. Cambridge: Cambridge University Press.
Beckwith, K. (2001). Gender frames and collective action: Configurations of masculinity in the Pittson Coal strike. *Politics & Society*, *29*(2), 297–330.
Benford, R. (1993). 'You could be the hundredth monkey': Collective action frames and vocabularies of motive within the nuclear disarmament movement. *Sociological Quarterly*, *34*, 195–216.
Benford, R., & Snow, D. (2000). Framing processes and social movements: An overview and assessment. *Annual Review of Sociology*, *26*, 611–639.
Bickerstaff, K., & Agyeman, J. (2009). Assembling justice spaces: The scalar politics of environmental justice in north-east England. *Antipode*, *41*(4), 781–806.
Bosco, F. (2001). Place, space, networks and the sustainability of collective action. *Global Networks*, *1*(4), 307–329.
Boykoff, J., & Laschever, E. (2011). The Tea Party Movement, framing and the US media. *Social Movement Studies*, *10*(4), 341–366.
Bulkeley, H., Carmin, J., Castan Broto, V., Edwards, G., & Fuller, S. (2013). Climate justice and global cities: Mapping emerging discourses. *Global Environmental Change*, *23*(5), 914–925
Bullard, R. (1999). Dismantling environmental racism in the USA. *Local Environment*, *4*(1), 5–19.
Caney, S. (2010). Climate change and the duties of the advantaged. *Critical Review of International Social and Political Philosophy*, *13*, 203–228.
Cantzler, J. (2007). Environmental justice and social power rhetoric in the moral battle over whaling. *Sociological Inquiry*, *77*(3), 483–512.
Caren, N., Ghoshal, R., & Ribas, V. (2011). A social movement generation: Cohort and period trends in protest attendance and petition signing. *American Sociological Review*, *76*(1), 125–151.
Chi, C. (2001). Capitalist expansion and indigenous land rights: Emerging environmental justice issues in Taiwan. *Asia Pacific Journal of Anthropology*, *2*(2), 135–153.
Cox, K. (1998). Spaces of dependence, spaces of engagement and the politics of scale. *Political Geography*, *17*(1), 1–23.
Cress, D., & Snow, D. (2000). The outcomes of homeless mobilization: The influence of organization, disruption, political mediation and framing. *American Journal of Sociology*, *105*(4), 1063–1104.
Cresswell, T. (2006). *On the move*. London: Routledge.
Davies, A. (2006). Environmental justice as subtext or omission: Examining discourses of anti-incineration campaigning in Ireland. *Geoforum*, *37*, 708–724.
Dawson, J. (2000). The two faces of environmental justice: Lessons from the eco-nationalist phenomenon. *Environmental Politics*, *9*(2), 22–60.
Debbane, A., & Keil, R. (2004). Multiple disconnections: Environmental justice and urban water in Canada and South Africa. *Space and Polity*, *8*, 209–225.
Delaney, D., & Leitner, H. (1997). The political construction of scale. *Political Geography*, *16*, 93–97.
Della Porta, D., & Mosca, L. (2007). *In movimento*: 'Contamination' in action and the Italian Global Justice Movement. *Global Networks*, *7*(1), 1–27.
Di Chiro, G. (2008). Living environmentalisms: Coalition politics, social reproduction, and environmental justice. *Environmental Politics*, *17*(2), 276–298.
Diani, M. (1996). Linking mobilization frames and political opportunities: Insights from regional populism in Italy. *American Sociological Review*, *61*, 1053–1069.
Diani, M. (2000). The relational deficit of ideologically structured action. *Mobilization: an International Journal*, *5*, 17–24.
Diani, M. (2005). Cities in the world: Local civil society and global issues in Britain. In D. Della Porta & S. Tarrow (Eds.), *Transnational protest and global activism: People, passions, and power* (pp. 45–67). Lanham, MD: Rowman and Littlefield.
Fan, M.-F. (2006). Environmental justice and nuclear waste conflicts in Taiwan. *Environmental Politics*, *15*(3), 417–434.

Franquemanque, G. (2007). From Larzac to the altermondialist mobilisation: Space in environmental movements. *Environmental Politics*, *16*(5), 826–843.

Fraser, N. (1997). *Justice interrupts: Critical reflections on the postsocialist conditions*. New York, NY: Routledge.

Fraser, N. (2008). *Scales of justice*. Cambridge: Polity Press.

Fraser, N., & Honneth, A. (2003). *Redistribution or recognition? A political–philosophical exchange*. London: Verso.

Gamson, W. (1992). *Talking politics*. Cambridge: Cambridge University Press.

Gamson, W., Fireman, B., & Rytina, S. (1982). *Encounters with an unjust authority*. Dorsey, IL: Homewood.

Gerhards, J., & Rucht, D. (1992). Mesomobilization: Organizing and framing in two protest campaigns in West Germany. *American Journal of Sociology*, *98*, 555–595.

Gibson-Wood, H., & Wakefield, S. (2013). Participation, white privilege and environmental justice. *Antipode*, *45*(3), 641–662.

Gieryn, T. (2000). A space for place in sociology. *Annual Review of Sociology*, *26*, 463–496.

Giugni, M. (2011). Political opportunity: Still a useful concept? In M. Hanagan & C. Tilly (Eds.), *Contention and trust in cities and states* (pp. 271–283). Dordrecht: Springer.

Goffman, E. (1974). *Frame analysis: An essay on the organization of the experience*. New York, NY: Harper Publishers.

Haarstad, H., & Floysand, A. (2007). Globalization and the power of rescaled narratives: A case of opposition to mining in Tambogrande. *Political Geography*, *26*, 289–308.

Harrison, J. (2006). 'Accidents' and invisibilities: Scaled discourse and the naturalization of regulatory neglect in California's pesticide drift conflict. *Political Geography*, *25*(5), 506–529.

Harvey, D. (1996). *Justice, nature and the geography of difference*. Cambridge: Blackwell.

Harvey, D. (2000). *Spaces of hope*. Berkeley, CA: University of California.

Heaney, M., & Rojas, F. (2006). The place of framing: Multiple audiences and anti-war protests near Fort Bragg. *Qualitative Sociology*, *29*, 485–505.

Heynen, N. (2003). The scalar production of injustice within the urban forest. *Antipode*, *35*, 980–998.

Hobson, K. (2006). Enacting environmental justice in Singapore: Performative justice and the Green Volunteer Network. *Geoforum*, *37*, 671–681.

Holifield, R. (2004). Neoliberalism and environmental justice in the United States environmental protection agency. *Geoforum*, *35*, 285–297.

Holifield, R. (2009). Actor-network theory as a critical approach to environmental justice. *Antipode*, *41*(4), 637–658.

Holifield, R. (2012). Environmental justice as recognition and participation in risk assessment: Negotiating and translating health risk at a superfund site in Indian country. *Annals of the Association of American Geographers*, *102*(3), 591–613.

Holifield, R., Porter, P., & Walker, G. (2009). Spaces of environmental justice: Frameworks for critical engagement. *Antipode*, *41*(4), 591–612.

Houston, D. (2013). Environmental justice storytelling: Angels and isotopes at Yucca Mountain, Nevada. *Antipode*, *45*(2), 417–435.

Houston, D., & Pullido, L. (2002). The work of performativity: Staging social justice at the University of Southern California. *Environmental and Planning D*, *20*(4), 401–424.

Ishiyama, N. (2003). Environmental justice and American Indian tribal sovereignty: Case study of a land-use conflict in Skull Valley, Utah. *Antipode*, *35*(1), 119–139.

Koopmans, R. (2004). Political opportunity structure: Some splitting to balance the lumping. In J. Goodwin & J. Jasper (Eds.), *Rethinking social movements* (pp. 61–73). Lanham, MD: Rowman and Littlefield.

Kurtz, H. (2003). Scale frames and counter-scale frames: Constructing the problem of environmental justice. *Political Geography*, *22*, 887–916.

Leitner, H. (1997). Reconfiguring the spatiality of power: The construction of a supranational migration framework for the European Union. *Political Geography*, *15*, 123–143.

Leitner, H. (2003). The politics of scale and networks of spatial connectivity: Transnational inter-urban networks and the rescaling of political governance in Europe. In S. Sheppard & B. McMaster (Eds.), *Scale and geographic inquiry: Nature, society and method*. Oxford: Blackwell Publishers.

Leitner, H., & Miller, B. (2007). Scale and the limitations of ontological debate: A commentary on Marston, Jones and Woodward. *Transactions of the Institute of British Geographers*, *32*, 116–125.

Leitner, H., Sheppard, E., & Sziarto, K. (2008). The spatialities of contentious politics. *Transactions of the Institute of British Geographers*, *33*, 157–172.

Lieshout, M., Dewulf, A., Aarts, N., & Termeer, C. (2011). Do scale frames matter? Scale frame mismatches in the decision-making process about a "mega farm" in a small Dutch village. *Ecology and Society*, *16*(1), 38–50.

Lievianos, R. (2012). Certainty, fairness and balance: State resonance and environmental justice policy implementation. *Sociological Forum*, *27*(2), 481–503.

Marston, S. (2000). The social construction of scale. *Progress in Human Geography*, *24*, 219–242.

Marston, S., Jones, J., & Woodward, K. (2005). Human geography without scale. *Transactions of the Institute of British Geographers*, *30*, 416–432.

Martin, A., Gross-Camp, N., Kebede, B., McGuire, S., & Munyarukaza, J. (2013). Whose environmental justice? Exploring local and global perspectives in a payments for ecosystem services scheme in Rwanda. *Geoforum*, *54*(1), 167–177.

Martin, D. (2003). Place-framing as place-making: Constituting a neighborhood for organizing and activism. *Annals of the Association of American Geographers*, *93*(3), 730–750.

Massey, D. (2005). *For space*. London: Sage.

McCauley, D. (2011). Bottom-up Europeanization exposed: Social movement theory and non-state actors in France. *Journal of Common Market Studies*, *49*(5), 1019–1042.

McCauley, D. (2013). Protest, politics and produce: A resource account of anti-GMO activism. *Local Environment*, *15*(1), 171–193.

McCauley, D., Heffron, R., Stephan, H., & Jenkins, K. (2013). Advancing energy justice: The triumvirate of justice tenets. *International Energy Law Review*, *107*(1), 1–9.

McDonald, D. (2002). *Environmental justice in South Africa*. Athens, OH: Ohio University Press.

McGammon, H., Muse, D., Newman, H., & Terrel, T. (2007). Movement framing and discursive opportunity structures. *American Sociological Review*, *72*, 723–749.

Moore, A. (2008). Rethinking scale as a geographical category: From analysis to practice. *Progress in Human Geography*, *32*(2), 203–225.

O'Brien, K. (2011). Responding to environmental change: A new age for human geography? *Progress in Human Geography*, *35*(4), 542–549.

Okereke, C. (2006). Global environmental sustainability: Intergenerational equity and conceptions of justice in multi-lateral environmental regimes. *Geoforum*, *37*, 725–738.

Okereke, C., & Dooley, K. (2010). Principles of justice in proposals and policy approaches to avoided deforestation. *Global Environmental Change*, *20*, 82–95.

Paasi, A. (2004). Place and region: Looking through the prism of scale. *Progress in Human Geography*, *28*, 536–546.

Pellow, N., & Brulle, R. (2005). *Power, justice, and the environment: A critical appraisal of the environmental justice movement*. London: MIT Press.

Powell, R. (2011). Frames and narratives as tools for recruiting and sustaining group members. *Sociological Inquiry*, *81*(4), 454–476.

Reed, M., & George, C. (2011). Where in the world is environmental justice? *Progress in Human Geography*, *35*(6), 835–842.

Reid, L., Sutton, P., & Hunter, C. (2010). Theorizing the meso level: The household as a crucible of pro-environmental behaviour. *Progress in Human Geography*, *34*(3), 309–327.

Routledge, P. (2003). Convergence space: Process geographies of grassroots globalisation networks. *Transactions of the Institute of British Geographers*, *28*, 333–349.

Schlosberg, D. (2004). Reconceiving environmental justice: Global movements and political theories. *Environmental Politics*, *13*(3), 517–540.
Schlosberg, D. (2007). *Defining environmental justice: Theories, movements and nature*. Oxford: Oxford University Press.
Schlosberg, D. (2013). Theorizing environmental justice: The expanding sphere of a discourse. *Environmental Politics*, *22*(1), 37–55.
Schlosberg, D., & Carruthers, D. (2010). Indigenous struggles, environmental justice and community capabilities. *Global Environmental Politics*, *10*(4), 12–35.
Sen, A. (2009). *The idea of justice*. London: Allen Lane.
Sewell, W. (2001). Space in contentious politics. In R. Aminzade, J. Goldstone, D. McAdam, E. Perry, W. Sewell, S. Tarrow, & C. Tilly (Eds.), *Silence and voice in contentious politics* (pp. 51–88). Cambridge: Cambridge University Press.
Sheppard, E. (2002). The spaces and times of globalization: Place, scale, networks, and positionality. *Economic Geography*, *78*(3), 307–330.
Shmueli, D. (2008). Framing in geographical analysis of environmental conflicts: Theory, methodology and three case studies. *Geoforum*, *39*, 2048–2061.
Snow, D. (2004). Framing processes, ideology and discursive fields. In D. Snow, S. Soule, & H. Kriesi (Eds.), *The Blackwell companion to social movements* (pp. 380–412). Oxford: Blackwell Publishing.
Snow, D., & Benford, R. (1988). Ideology, frame resonance, and participant mobilization. *International Social Movement Research*, *1*, 197–217.
Snow, D., & Benford, R. (1992). Master frames and cycles of protest. In A. Morris & C. Mueller (Eds.), *Frontiers in social movement theory* (pp. 132–151). New Haven, CT: Yale Press.
Szasz, A. (1994). *Ecopopulism: Toxic waste and the movement for environmental justice*. Minneapolis, MN: University of Minnesota Press.
Sze, J., & London, J. (2008). Environmental justice at the crossroads. *Sociological Compass*, *2*(4), 1331–1354.
Sze, J., London, J., Stirling, F., Gambirazzio, G., Filan, T., & Cardenasso, M. (2009). Defining and contesting environmental justice. *Antipode*, *41*(4), 807–843.
Tarrow, S. (2005). The dualities of transnational contention. *Mobilizations*, *10*, 53–72.
Tarrow, S., & della Porta, D. (2005). Globalization, complex internationalism, and transnational contention. In D. della Porta & S. Tarrow (Eds.), *Transnational protest and global activism* (pp. 227–246). Lanham, MD: Rowman and Littlefield.
Tarrow, S., & McAdam, A. (2004). Scale shift in transnational contention. In D. della Porta & S. Tarrow (Eds.), *Transnational movements and global activism* (pp. 121–150). Lanham, MD: Rowan and Littlefield.
Taylor, D. (2000). The rise of the environmental justice paradigm: Injustice framing and the social construction of environmental discourses. *American Behavioral Scientist*, *43*(4), 508–580.
Tilly, C. (1986). *The contentious French*. Cambridge, MA: Harvard University Press.
Tilly, C. (1995). *Popular contention in Great Britain*. Cambridge, MA: Harvard University Press.
Towers, G. (2000). Applying the political geography of scale: Grassroots strategies and environmental justice. *Professional Geographer*, *52*, 23–36.
Urkidi, L., & Walter, M. (2011). Dimensions of environmental justice in anti-gold mining movements in Latin America. *Geoforum*, *42*, 683–695.
Verstraete, G., & Cresswell, T. (2002). *Theorizing place, placing mobility: The politics of representation in a globalized world*. Amsterdam: Rodopi.
Visser, G. (2003). Spatialities of social justice and local government transition: Notes on and for a South African social justice discourse. *South African Geographical Journal*, *85*(2), 99–111.
Walker, G. (2009a). Beyond distribution and proximity: Exploring the multiple spatialities of environmental justice. *Antipode*, *41*(4), 614–636.
Walker, G. (2009b). Globalizing environmental justice. *Global Social Policy*, *9*(3), 355–382.
Walker, G. (2012). *Environmental justice: Concepts, evidence and politics*. Abingdon: Routledge.

Walker, G., & Day, R. (2012). Fuel poverty as injustice: Integrating distribution, recognition and procedure in the struggle for affordable warmth. *Energy Policy*, *49*, 69–75.

Wang, D., & Soule, S. (2012). Social movement organizational collaboration: Networks of learning and the diffusion of protest tactics. *American Journal of Sociology*, *117*(6), 1674–1722.

Williams, G., & Mawdsley, E. (2006). Postcolonial environmental justice: Government and governance in India. *Geoforum*, *37*, 660–670.

Williams, R. (1999). Environmental injustice in America and its politics of scale. *Political Geography*, *18*(1), 49–73.

Wright, R., & Boudet, H. (2012). To act or not to act: Context, capability, and community response to environmental risk. *American Journal of Sociology*, *118*(3), 728–777.

Zald, M. (2000). Ideologically structured action: An enlarged agenda for social movement research. *Mobilization: An International Journal*, *5*(1), 1–16.

ANTHROPOLOGY AT THE RED-GREEN CROSSROADS

Brian McKenna

ABSTRACT

This chapter will examine ideological debates currently taking place in academics. Anthropologists – and all academic workers – are at a crossroads. They must determine what it means to "green the academy" in an era of permanent war, "green capitalism," and the neoliberal university (Sullivan, 2010). As Victor Wallis makes clear, "no serious observer now denies the severity of the environmental crisis, but it is still not widely recognized as a capitalist crisis, that is, as a crisis arising from and perpetuated by the rule of capital, and hence incapable of resolution within the capitalist framework."

Keywords: Anthropology; left-wing; environmental theory; ideology; analysis; debates

> Not everything that is faced can be changed, but nothing can be changed until it is faced.
> – James BaldwinMcKenna & Darder, 2011

Amidst the terror of neoliberalism (Bauman, 2006; Collins, Leonardo, & Williams, 2008; Giroux, 2004, 2011b; Harvey, 2010) world-historic catastrophes unfold, from the financial meltdown of 2008 to the election of Donald Trump in 2016. This authoritarian administration, knee deep in Climate Change denial, is intent on dismantling the social state by using its regal powers of command, control, and propaganda to enhance capital accumulation on a world

Environmental Criminology: Spatial Analysis and Regional Issues
Advances in Sustainability and Environmental Justice, Volume 20, 21–41

ISSN: 2051-5030/doi:10.1108/S2051-503020170000020002

scale. They are working hard to roll back the 1960s, including environmental protections, subverting the EPA to allow oil, gas, and fracking interests proceed at will with as few regulations as possible. U.S. citizens, long subdued by illusion, spectacle, and fear, are slowly growing conscious that they are faced with a class war, a real one. The greatest threat to Homeland Security is internal. It's the Homeland Security state, a vast military–industrial–financial complex, armed with the most sophisticated surveillance technology in world history, led by a proto-fascist administration which eagerly does their bidding. Neoliberalism (capitalism with the gloves off) is transforming into something much darker and perverse.

The Iraq War forebode these traumas. The Nuremberg Protocols were ignored (Mandel, 2005) in a mad rush for oil as the United States invaded Iraq under false pretenses. Tens of thousands died. Millions were scarred for life as the United States devolved into a torture state (Giroux, 2011a). It was a "public engagement" from above, spawned by neoconservatives and strongly supported by a reliable friend: the academy.

Several U.S. universities were active participants in the war including my alma mater, Michigan State University. Peter McPherson, MSU's President, was recruited by President Bush to oversee Iraq's economic restructuring in 2003. Standing with the imperial rulers of an occupying army, McPherson (the former head of USAID) worked to impose his political and economic vision on a captive nation (McKenna, 2003a, 2003b).

"If you don't do enough to create a political constituency for privatization now," he told Fortune Magazine, "then it will get killed in the cradle" (Kahn, 2003; McKenna, 2003a, 2003b).

For his free market zeal, one of McPherson's own team members accused him of believing in an "ideological nirvana," according to Kahn (Kahn, 2003).

Back home on MSU's campus, there were no public protests or teach-ins about the war nor about the activities of MSU's President. McPherson returned home, after a six month leave of absence, to a hero's welcome. Tenured faculty were, by and large, publicly silent.

I worked as a local environmental journalist (and applied anthropologist) at the time, 2001–2004, and wrote several articles that connected McPherson's neoliberal work abroad with his neoliberal work at home, focusing on his anti-environment activities. When his good friend Vice president Cheney came to campus to deliver the graduation MSU address in 2002, I wrote a column, "*King McPherson's New Throne*" (2002) in which I reported McPherson's recent appointment to chair the Department of Energy's powerful external advisory committee and asked him to help release 15,000 pages of minutes from Cheney's secret energy industry meetings.

Later, when the MSU President served in Iraq, I wrote "*Who is Michigan's Empire Man?*" (2003) and catalogued a series of his anti-green activities. McPherson energetically supported General Motors in intimidating Lansing's Westside citizens (encompassing about 4,000 households) by pressuring them – as part of a Lansing State Journal petition to which he lent his signature – not to file an air pollution appeal as was their right under the Clean Air Act (McKenna, 2002a); he agreed to let MSU police infiltrate a registered campus student group, United Students Against Sweatshops (since renamed Students for Economic Justice), who were concerned about the World Trade Organization and unfair labor practices; and he refused to abide by a democratic MSU student referendum in March 2003 that supported a proposed $5-per-semester tax to pay for renewable energy on campus. These connections "could have been a teachable green moment of epic proportions for the faculty" (McKenna, 2003a, 2003b). But was not.

GREENING THE CAMPUS

In the midst of the United States' world-wide dispossession, a burgeoning movement seeks to "green the campus," a development that could theoretically challenge the corporatization of the university and help reclaim democratic life (Barlett & Chase, 2004; Sullivan, 2010). Across "new landscapes of inequality" (Collins et al., 2008), colleges and universities are awash in civic engagement initiatives, public–private partnerships, and campus sustainability programs. The American Association for Sustainability in Higher Education founded in 2006 defines sustainability "in an inclusive way, encompassing human and ecological health, social justice, secure livelihoods, and a better world for all generations." These efforts could seriously challenge the status quo.

And yet, at MSU, and at campuses around the country, the levels of critical civic engagement about "sustainability" is a pittance of what it could be. Anthropologist Peggy Barlett is at the forefront of the movement and helped found Emory University's Office of Sustainability in Atlanta. She identifies a paradox in a profession, anthropology, that it honors holism in theory but defers from holism where they work: the campus. "For … anthropologists [our] loyalty [is] to the place we did our fieldwork," she said, "we're experts on Papua New Guinea or sophisticated about Paris. There's no prestige in being knowledgeable about the particular town or bioregion where our university may be located. This attitude transmits to our students a notion that, well you don't have to know anything about where you are to be a successful person and a knowledgeable citizen. In this particular era, that's a dangerous habit" (Wells, 2010).

Anthropologists – and all academic workers – are at a crossroads. They must determine what it means to "green the academy" in an era of permanent war, "green capitalism," and the neoliberal university (Sullivan, 2010). As Victor Wallis makes clear, "no serious observer now denies the severity of the environmental crisis, but it is still not widely recognized as a capitalist crisis, that is, as a crisis arising from and perpetuated by the rule of capital, and hence incapable of resolution within the capitalist framework" (Wallis, 2010).

Anthropologist Merrill Singer concurs, calling for "an eco-nomic transcendence of the capitalism mode of production" (Singer, 2010, p. 128).

A PLUNDERING WORLD

"When practiced properly," David Price reminds us, "anthropology is a threatening science" (Price, 2004, p. 29). Anthropologist Laura Nader concurs, plunder, she says, is a chief dynamic of our times. The Latin American novelist Eduardo Galeano captures the zeitgeist. "Plunder, internal and external, was the most important means of primitive accumulation of capital, an accumulation which, after the Middle Ages, made possible a new historical stage in world economic evolution" (Mattei & Nader, 2008). Primitive accumulation – wars, violence, enclosures, and privatization – is a chief means by which capitalism appropriates the commons.

Primitive accumulation and ethnocide are ongoing. This is especially evident in one of my focal areas: Indians of North America (McKenna, 2006, 2011a, 2011b). The Bullfrog Film *Homeland: Four Portraits of Native Resistance* (2005) illustrates the continued destruction of the Indian commons (Biolsi, 2004; Oswalt, 2009). Gail Small, a member of the Northern Cheyenne Indian Tribe in Lame Deer, is profiled. "You put in 75,000 methane gas wells around our reservation, you take our ground water, pollute our air, destroy our rivers, the Cheyenne here will probably not be able to survive. We'll have a wasteland here. That's what's at stake here. Where will the Cheyenne go?"

Where indeed? Human beings are treated as so much waste in this age of disposability (Giroux, 2010). The earth has become the *private* property of a human cabal. Can reason and democracy alter the neoliberal capitalist tide and rescue the commons?

This chapter surveys the wide breath of green anthropological activity, both theoretical and applied, in ecological and environmental realms (e.g., conservation, biodiversity, ethnobotany, water security, and eco-cities). The amount of material is vast, so much so that Les Sponsel, a leader in the field, exclaims, "no one can keep track of it all."

I focus on four pressing areas of critical green practice in anthropology with which I am involved: backyard anthropology (Johnston, 2001), critical public pedagogy (Sandlin, Schultz, & Burdick, 2009), activist-applied anthropology, and the political ecology of health.

A RED-GREEN COMMONS?

Universities are a kind of commons, an essential bulwark for creating an alert democracy to address these monumental social problems. But today, as indicated above, universities are fast becoming capitalist knowledge factories

(Aronowitz, 2000), a central tier of Eisenhower's feared "military industrial academic complex" (Giroux, 2007).

Can greening the academy help reverse the neoliberal cataclysm? Will it educate for "revolutions of justice?" Anthropology asks these questions in its theories, ethnographies, and classroom curricula. As such, its fundamental green message is radical (Anderson, 1996; Johnston, 2007; Singer, 2010). Anthropology charts the long view of our species detailing the prehistory of primitive communism (constituting vast millennia of species history) to primitive accumulation which is our lot today (Merchant, 2005).

Every semester I show the film *N!ai, The Story of a !Kung Woman* in my Introduction to Anthropology class. Filmed by ethnographer John Marshall, on and off over 28 years, the film documents the hunter and gatherer Bushmen of Botswana, who only required 2–3 days per week for subsistence procurement. They ate 105 plant species from the local ecology which constituted 60–70% of their diet (the mongongo fruit and nut being a staple) and also hunted an occasional giraffe for most of their 20,000-year-old adaptation (and transformation) of their environment. The film then portrays the Kung San's colonization as they were cast onto reservations, stricken with tuberculosis, and pressured into the South African Army of the apartheid era. In just one generation under the South Africans they were victims of ethnocide and some would argue genocide (Robbins, 2008).

Afterwards I discuss the work of Marshall Sahlins who in his classic *Stone Age Economics* (Sahlins, 1972) calls the Kung's 20,000 year type of social organization "the original affluent society." I contrast Kung society with American students' everyday lives, mired in student debt, part-time jobs, and unknown futures.

"There it is," I say, "Primitive communism versus primitive accumulation, all in one hour of viewing." "Who are the primitives?" I ask.

How green is anthropology? What gains have been made? What limitations constrict its movement and application to real world scenarios? What is my vision? I address each of these in this chapter. Indeed the gains are enormous.

In 1996 the Anthropology and Environment section of the AAA (http://www.eanth.org/) was established to catalogue the gains and advance green anthropology. Anthropologist Barbara Johnston was a key activist in forming the subdiscipline, working as "disciplinary organizer," serving on task forces, committees, writing reports, booklets, and books, responding to community demand for this work. Its mission, is, in part, to "coordinate a discipline-wide collaborative effort that would renew anthropology's commitment to holism, to reach beyond our own field to incorporate research and practitioners from other disciplines," according to Johnston. It has not come close to accomplishing that high standard, however it is a valuable group with 678 members (as of June 2011). It has journals (*Human Ecology, Journal of Ecological Anthropology, Ecological and Environmental Anthropology*), anthologies and textbooks, publisher's series, specialists, activists, programs, and courses; and an excellent listserv (see above website for access). It has searchable archives which is an indispensible scholarly resource. As of June 2011 the EANTH list had 1,513 subscribers.

But three key questions loom. First, how does anthropology, ostensibly a science of holism, reconcile the overspecialized practice it has largely become (Harvey, 2009) with the widespread call to speak plainly for the educated lay public beyond academia? (Eriksen, 2006). Second, how does anthropology reconcile its limiting context within the "military-industrial-academic complex" on the one hand with the liberating movement to radically green the academy – and the country – on the other? (Johnston, 2007). And third, how does anthropology situate itself with respect to a failed neoliberal ideology – in which capitalism has proven unsustainable – given anthropology's long-standing commitment to defend indigenous peoples, democracy, and the commons? (Singer, 2010).

ANTHROPOLOGY HAS BECOME A "SOCIAL SILENCE"

The last person to write a chapter describing "green anthropology" for a general readership was William Balee, in 1996. In *Greening the College Curriculum* (Balee, 1996), Balee's job was to discuss the "vast subject matter" of environmental/ecological anthropology up to that time. Befitting his own expertise, Balee focused on his specialty, forest ecology.

"Jungle is as much a myth as Tarzan is," Balee explained. "The ecologist who tries to study undisturbed communities (such as a virgin forest) is likely to spend his whole life trying to find one!" These aphoristic statements draw attention to the "pristine myth," which has served as justification for colonization of North America (Denevan, 1996). Perhaps no greater case of normative blindness (i.e., how culture renders the obvious invisible) exists in America than Cahokia, the third largest structure in pre-capitalist North America. It was a city of about 20,000 people in a region that had over 50,000. The Mississippian peoples had departed around 1,300, probably after an ecological catastrophe, but left behind nearly a 100 great structures including Monks Mound, at 100 feet high and 785,000 cubic yards, the largest pyramid north of Mexico. It took over 200 years to build. Existing today on the outskirts of St. Louis, Cahokia, was seen as empty land with the mounds left behind by the Lost Tribes of Israel. Certainly Indians could not have created it! Manifest Destiny required this view (Pauketat, 2009).

One would think that this is pretty significant scholarship that deserved a wider distribution. That's what Charles Mann (2005) thought after hearing Balee talk "about anthropogenic forests – forests created by Indians centuries or millennia in the past." Said Mann, "it was a concept I'd never heard of before." He reflected, "Gee, someone ought to put all this stuff together,

I thought. It would make a fascinating book." He waited but nothing happened. So he finally wrote it. His resulting book, *1491* became a national bestseller (Mann, 2005).

Why didn't Balee write that book? In fact, it is a common story in anthropology: a lack of urgency to communicate with the public. This has become a crisis. The 2009 American Anthropology Association annual meeting was titled, "The End/s of Anthropology," expressing worry about the profession's future. Unlike other disciplines, anthropology has not produced many public intellectuals. It suffers from the challenge of "public engagement" and is groping for strategies, models, and ideas. The conference asked its 12,000 members "What is the relevance of anthropology in today's world?" The relevance is enormous, particularly its "green" contributions. But relatively few academics, citizens, nor even anthropologists are aware of the scope. One reason is that anthropology has become a "social siLence" (McKenna, 2011a, 2011b; Tett, 2009). Anthropologist Gillian Tett, the Managing Director of the Financial Times, said as much in a blistering speech at the 2010 AAA meetings (McKenna, 2011a, 2011b). The author of the bestselling *Fool's Gold* (2009) which recounts how she predicted the 2008 meltdown, argues that anthropologists need to project themselves more forcefully into the culture. "Anthropologists are well trained to absorb information, not project it. They have to emit."

Public pedagogy and journalism are not new ideas in anthropology (Eriksen, 2006). Franz Boas, Margaret Mead, Ashley Montague, Louis Leaky, Marvin Harris, and more recently David Mayberry Lewis, producer of the Millennium book (1992) and TV Series on PBS were active public communicators. Boas, one of the founders of U.S. anthropology, was adamant about marshaling civic voice against injustice. He developed the methodology of historical particularism (i.e., that cultures are to be studied diachronically and in depth) and was a tireless advocate for "cultural relativity" (i.e., that cultures should be analyzed on their own terms). He became an outspoken critic of racism. As Price describes it, "Boas and his students saw it as their duty to bring this perspective to the streets, newspapers, radio broadcasts, and government agencies" (Price, 2004, p. 34).

What happened then? Read on.

THE TREE OF KNOWLEDGE HAS TOO MANY BRANCHES

There are four classic branches within anthropology: physical (or biological), cultural, linguistics, and archaeology. Today applied anthropology is usually added as the "fifth field." Anthropologists are required to study the basics of the first four areas in graduate programs. All five branches have abiding environmental/ecological foci, and yet they have generally become isolated from one another.

Even within ecological/environmental anthropology there are divisions, subfields, and splits. There are four main approaches within "ecological anthropology," including cultural ecology, historical ecology, political ecology, and spiritual ecology, which developed historically in that order. Within these groupings there are a number of topical subfields and/or methods including ethnobotany, medical ecology, primate ecology, systems ecology, ethnoecology, behavioral ecology, information ecology, human ecology, forest ecology, aquatic ecology, arctic ecology, and more. Other evolving subfields from across anthropology as a whole include, prehistoric ecology, bioarcheaology, cognitive anthropology, Darwinian evolutionary ecology, and the political ecology of health. A special note must be made about scholars who pursue ecological approaches within the broad field of American Indian and Indigenous studies.

Let's look at the five major branches in order to consider the interstices of each of the various anthropological approaches and subfields. This allows us to see where ecological/environmental approaches to anthropology can be observed.

First, a word about terminology: Sponsel defines ecological anthropology as the exploration of how culture influences the dynamic interactions between human populations and ecosystems in their habitat through time. He views environmental anthropology as an applied dimension of ecological anthropology; others like Michael Dove prefer to see it as a more recent development replacing ecological anthropology.

For an excellent overview of the subdiscipline, with references, consult Sponsel's webpage (http://www.soc.hawaii.edu/Sponsel/).

From a *biological (or physical) anthropology* perspective, humans are "green" in their very being. We are all extensions of the Earth. We are culture bearing animals, descended from the trees, the result of biocultural evolution. That is, humans are nature, albeit a nature that we have had a hand in sculpting. Our large craniums (in juxtaposition to our Australopithecine ancestors) are dialectically related to the freeing of our hands and subsequent tool use. Our species' history is borne of material struggle, we now have the capacity to make meaning, learn, plan, and cooperate. We both adapt to and transform our environments, culturally, biologically, and politically. Our black, white, red, and brown skin colors are related to the latitudes we've traversed in and out of Africa beginning over a hundred thousand years ago.

Cultural anthropology is about making connections, revealing hidden assumptions, teaching about human diversity and subaltern cultures, uncovering origins (of humans, the state, one's landscape, oneself), and resisting oppression. There are a number of outstanding ecological ethnographies. One is *Priests and Programmers* by Stephen Lansing (1991) which argues that temple priests in Bali regulated irrigation networks for the rice farming system based on cooperative spiritual principles. These were undermined by the Western

Green Revolution (Nazarea, 2006). Another important work is *Against the Grain* (2008) an edited volume in honor of Pete Vayda. His methods of "event ethnography," and the "actor centered approach" are in many ways similar to the methods of investigative journalists. Tim Ingold is widely touted in the field for important scholarly contributions like *The Perception of the Environment* (2000) in which he theorizes about the ethnography of everyday pedestrian movements and how something as simple as walking binds time and place in people's experience, relationships, and life-histories. Ingold was influenced by anthropologist Gregory Bateson in books like *Steps to an Ecology of Mind* (1972). Bateson posited the mind as an immanent part of the whole interconnected system of organism–environment relations not as something confined in our bodies against nature "out there." Bateson left a mark on ecological anthropology influencing many, including Roy Rappaport and E. N. Anderson who, in books like *Ecologies of the Heart* (1996), argue that the emotional side of humanity is more basic than the mind. There are a number of excellent readers that span the field including Dove and Carpenter's *Environmental Anthropology: A Historical Reader* (2005) and Carrier and West's *Virtualism, Governance and Practice: Vision and Execution in Environmental Conservation* (2009). There is a debate within anthropology concerning the degree to which anthropologists ought to advocate for "ecological rights" versus those who think that they shouldn't. I take up this issue later, below.

Archaeology reconstructs the cultures and environments of prehistoric peoples through analyses of artifacts, fossils, natural, and material remains. Bioarchaeologists and paleopathologists can tell us much about prehistoric health, nutrition, environments, and even inequality based on investigation of remains like plants, teeth, skeletons, and coprolites (hardened human feces). Archaeologists are the ones who provide evidence for how civilizations in ancient Sumer and the American Southwest failed due to ecological destruction. Charles Redman was a trailblazer in this realm with his *Perspectives on Southwestern Prehistory* (1991). For an excellent textbook on this area, I refer readers to Hornborg and Crumley's *The World System and The Earth System: Global Socio-Environmental Change and Sustainability Since the Neolithic* (2006). Another valuable text is Crabtree and Campana's *Exploring Prehistory: How Archaeology Reveals our past* (2005). See reference section for American Anthropological Association link to best archaeology and environmental student Fieldschools (AAA, 2011).

Linguistic anthropology investigates how people classify local flora, fauna, and landscapes. It captures the referential contexts for how people explain illness symptoms, medicine, herbology, and treatments. This has enormous theoretical and practical implications. Michael Halliday was a founder of social semiotics who has had a profound impact on ethnolinguistics. For example, his analysis of "economic growth," described how its English language metaphoric association with notions of large, tall, and good provides "growth with a positive connotation when in fact it has severely negative ecological consequences"

(Jones, 2010). Berlin and Berlin (1996) collected ethnobotanical data over seven years to reveal Tzeltal and Tzotzil Maya's in-depth knowledge of gastronomic illness and treatment. A recent article in the *Annual Review of Anthropology* summarizes the methodologies used to analyze environmental discourses including discourse analysis, ecolinguistics, and ecocritical linguistics (Mühlhäusler & Peace, 2006). Eugene Hunn's *A Zapotec Natural History: Trees, Herbs, and Flowers, Birds, Beasts, and Bugs in the Life of San Juan Gbëë* (2008) is an excellent portrait of the linguistic power of an Zapotec Indian community located in the state of Oaxaca. Anthropologist Daniel Moerman constructed the world's most complete inventory of Native American Ethnobotany. The searchable database is available at http://herb.umd.umich.edu/. Moerman devoted 25 years to the task of gathering together the accumulated ethnobotanical knowledge on more than 4,000 plants. More than 44,000 uses for these plants by various tribes are documented here. The database contains thousands of details on foods, drugs, dyes, and fibers of Native American Peoples, derived from plants. Moerman is an Emeritus Professor from my campus, the University of Michigan-Dearborn. One of his former students developed a brochure identifying Native American plants on campus. The Environmental Interpretive Center sponsors walks based on it, as do I.

The fifth discipline of *applied anthropology* has grown exponentially since academic jobs have become scarce. One could make the argument that most anthropologists referenced in this chapter do applied work to one degree or another and indeed, there are rampant debates within the discipline about what constitutes "applied" or public work (Ervin, 2005). The *Society for Applied Anthropology Newsletter* under the editorship of Tim Wallace is a quarterly publication that provides timely articles on this valuable work (http://www.sfaa.net/newsletter/newsletter.html). A leader in environmental applied anthropology is Barbara Johnston. She works as a tireless bridgewalker between NGOs, citizen groups, governments, and the academy. Her new book *Life and Death Matters: Human Rights, Environment and Social Justice (2011, 2nd edition)* is essential reading (Johnston, 1994). In the wake of Japan's Fukushima nuclear disaster, Johnston has written articles for the *Bulletin of the Atomic Scientists, CounterPunch, Truth-out.org*, the SFAA Newsletter Johnston (May, 2011) as well as appearing on television. Another bridgewalker is David Cassagrande who mines the interstices between ecological theory, human cognition, and public policy. His groundbreaking work with interdisciplinary teams on medical botany among the Tzeltal Maya and rural communities on the Mississippi floodplain demonstrate that community resilience is more determined by equal distribution of information rather than the complexity of information (Cassagrande et al., 2007). I have worked as a "data democratizer" myself a medical school evaluator, Executive Director of the environmental organization LocalMotion ("better health through fewer toxins"), and public health worker. As a government environmental health official, I conducted an environmental health assessment for county government in Mid-Michigan, a 300-page work on water, air, and food that was eventually suppressed. Later it was released by Public Employees for Environmental Responsibility (McKenna, 2002a, 2002b; PEER, 2001).

OVERCOMING SOCIAL AMNESIA: HISTORY IS A WEAPON

In his "green curriculum" article Balee cited three major figures in the history of ecological anthropology, Julian Steward, Roy Rappaport, and Eric Wolf. Steward is a seminal figure who is important for breaking with armchair evolutionary theorists and inventing the method of "cultural ecology" (a term he coined) in his work with the Shoshoni Indians in the American Southwest (Steward, 1955). He investigated how a culture adapted and achieved what he called "homeostasis" in a given ecology through ones material culture (pots, tools, baskets, etc.). Rappaport advanced ecological theory in *Pigs for the Ancestors* (1968) where he argued that the Tsembaga Maring of Papua New Guinea regulated their culture through the ritual slaughter of pigs (McKenna, 2009).

Later Steward's student Eric Wolf broke with Steward and the systems theory of Rappaport. Wolf argued that cultural ecology had to become synthesized with political economy in order to account for the dialectical relationships between global and local forces. He critiqued Steward's notions of an isolated, bounded population as the unit of analysis and critiqued ideas that humans primarily seek adaptations to environments to achieve homeostasis. Wolf helped to establish the field of "political ecology." He asked, famously, "If there are connections everywhere why do we persist in turning dynamic, interconnected phenomena into static, disconnected things?"

In this light, other connections are relevant. Steward served as an expert witness for many years with the Department of Justice. Laura Nader charges that "Steward's social evolutionary theory became a matter of legal and political significance, a theory that legitimized the denial of indigenous rights to collectively hold lands" (Mattei & Nader, 2008, p. 104). By depicting the Northern Paiutes as too primitive to be organized (thus going against the Boasian tradition) he assisted the state in depriving them of lands, she said. Price argues that Steward self-censored much of his Marxist theory because it was dangerous at the time (Price, 2004). One influential scholar who openly did build on Marx was Marvin Harris whose *Rise and Anthropological theory* (2001 [1969]) and *Cultural Materialism* (2001 [1979]) are fundamental readings still.

Ecological anthropology has advanced tremendously since Steward. Much of the groundwork that supported the creation of the subdiscipline, in 1996, took place at the epoch changing United Nation's Rio de Janeiro Conference on the Environment (Rio Earth Summit) in 1992. Many anthropologists were active participants including a large contingent who advocated for indigenous rights issues. Darrell Posey organized a parallel event at the time called the Earth Parliament, a fifteen-day assembly of indigenous and minority groups. The strategy worked, influencing the main assembly to forge an agreement at the Rio Summit to "not carry out any activities on the lands of indigenous peoples that would cause environmental degradation or that would be culturally inappropriate." Many anthropological luminaries were involved in the two-week proceedings including Linda Raben, a former member of the AAA Committee for Human Rights. She worked at the Rainforest Foundation at the time. Another was Mexican anthropologist Lourdes Arizpe, the former Director of the Institute of Anthropological Research at the National University of Mexico (a major environmental anthropology prize is named after her). Others included, Terrence Turner and Emilio Moran (2010) who represented the AAA's environmental task force established by former AAA President Roy Rappaport. Mary Elmendorf, a winner of the prestigious Malinowski Award by the Society for Applied Anthropology attended as a delegate. Elmendorf was a recipient of the Nobel Peace Prize (along with CARE) and the person primarily responsible for educating the World Health Organization to address women's issues.

Anthropology may have produced legions more of these activist environmental anthropologists, were it not for the encroachments of the U.S. government. In a magnificent work of scholarship anthropologist Davis Price drew on more than 30,000 pages of FBI and government memoranda released under the Freedom of Information Act which described dozens of activist anthropologists who were prosecuted during the Red Scares of the 1940s and 1950s (Price, 2004). Price showed that it was not Communist Party membership or Marxist beliefs that attracted the most scrutiny but the levels of public action, particularly around racial justice. He argues, "McCarthyism took a large chunk out of American anthropology – a chunk so deep it continues to affect and limit the scope and approach of anthropology today" (Price, 2004, p. 33).

This may be a factor why few anthropologists are familiar with the work of political ecologist James O'Connor. Anthropologist Barbara Johnston worked closely with O'Connor in forming the influential environmental journal *Capitalism, Nature, Socialism*. In O'Conner's powerful and prescient book *Natural Causes, Essays in Ecological Marxism* (Guilford Press, 1998) he argues that the dynamics of capital revolve around not one, but two central contradictions. The first involves the traditional understanding of the contradiction between capitalism's productive forces and its productive relations. Capitalism's relentless motions to accumulate and "grow" come up against the necessary contradictions of overproduction, underproduction, and worker unrest. The second contradiction of capital addresses the ecological crises of our times. It arises from the way capital limits itself by impairing its own social and environmental conditions, using nature as both "tap and sink." O'Connor is tough on traditional Marxists for undertheorizing the role of ecology. But he is equally tough on most (liberal) ecologists for not reading enough political economy and Marx.

Similarly, in this age of privatization, one would expect that more environmentalists would be familiar with the groundbreaking work of anthropologist Bonnie McCay. Her research refutes the underpinnings of Garret Hardin's popular Malthusian thesis in "Tragedy of the Commons." Hardin, an ecologist, promoted a very pessimistic view of human nature, one ideologically supportive of capitalism. In analyzing an Old English commons Hardin asserted that each individual herdsman would selfishly add more cows to his herd while overgrazing the commons whose grass he would take for "free." Since each herdsman would do this, overgrazing resulted and the commons was damaged or destroyed. McCay drew upon her ethnographic fieldwork with Newfoundland fishing communities to demonstrate how they successfully managed common-pool resources in sustainable ways. She and her colleagues in books like "The Question of the Commons" (1987) and Enclosing the Commons (2002) document how communities can act as stewards with broader social interests beyond self-serving interests. "Common property" does not mean equality of access. It means systems of obligations, rights, and duties to protect shared resources.

When it comes to the commons, few places seem more obvious a place to work together than the growing campus sustainability movement. It is a "common ground." A leader in this movement is Peggy Barlett, an anthropologist at Emory University. Barlett conducts Sustainability Across the Curriculum Workshops for the Association for the Advancement of Sustainability in Higher Education. She has greatly advanced the theory and practice of the greening movement on campus (Wells, 2010).

A number of anthropologists are active in this bridgewalking work. In fact the two universities who have created formal schools of sustainability (Arizona State University and the University of South Florida) did so with the drive of anthropologists (Wells, 2010).

Carl Maida is a partner in a community-based participatory research project, Community Action for a Renewed Environment (CARE) Program – Reducing Toxic Risks, funded by the U.S. Environmental Protection Agency. His recent book, *Sustainability and Communities of Place* (2007) explores sustainable development as a local practice, world-wide. For the past two decades anthropologist Sam Beck (2009) has directed the New York City Urban Semester Program at Cornell University where he connects students' medical experiences with diverse urban ecological issues.

There is continued debate about the best approaches for campus involvement. Should anthropologists, in the words of one campus organizer, "simply tell the 'truth' or let the 'truth' dazzle gradually?" (McKenna & Darder, 2011). A leading anthropologist in the sustainable campus movement takes the latter view telling me that anthropologists often refrain from getting involved in sustainable campus initiatives "because we're disciplinarily trained to be very critical." She argued that "That stance is bad for forging good working relationships with non-academic partners, the heart of culture change work. And it makes it hard to accept the messy realities of compromise. Power structures are only changed in incremental struggles that are not fun and require us to give up control. Anthropologists are not so good at long-term dialogue with leaders to help them see a different way, let alone organizing to shift the legal or grassroots structures that bind corporate (or academic) leaders."

Other anthropologists take a more urgent view of social change strategies, reflecting "the 'truth' should thunder mightily" approach. See more on this below.

Thousands of ecological/environmental anthropologists wrestle with these issues of public voice. They are involved in conservation, ecotourism, sustainable urban development, sustainable agriculture, sustainable forestry, and sustainable fisheries initiatives around the world. In a 2011 message to EANTH listserv nine leading environmental anthropologists, including its current President, Paige West, provided links to their websites to improve public accessibility about their applied work. I provide links in the reference section to each one: Dan Brockington, University of Manchester (2011); Bram Buscher (2011); Wolfram Dressler, University of Queensland (2011); Rosaleen Duffy, University of Manchester (2011); Robert Fletcher, National Peace Academy (2011); Jim Igoe, Dartmouth College, Katja Neves, Concordia University (2011); Sian Sullivan, Birbeck College (2011); and Paige West, Columbia University (2006).

SILO-BUSTING

Medical doctors tend to ignore the environmental etiologies to illness and disease. Are anthropologists reproducing this divide? Cecil Helman, in the most popularly read medical anthropology textbook, neglected environmental health for years. His recent 2007 book expounds on environmental issues but presents them mostly as an "add on" (i.e., he does not integrate them well with clinical medicine) largely reproducing the restricted clinical gaze of biomedicine (McKenna, 2008a, 2008b).

For over three decades Ann McElroy and Patricia Townsend have bridged disparate fields by developing the concept of "medical ecology." In 1977, they noted that no one had "pull[ed] all this work together in a coherent field of study," bridging disparate fields. The fifth of their book Medical Ecology in Ecological Perspective (2009) is an exceptional compendium of wide ranging literature and is extremely well written. It should be required reading for all anthropologists.

In a similar fashion, critical medical anthropologists Merrill Singer and Hans Baer have begun to integrate more environmental literature into their work, including *Global Warming and the Political Ecology of Health* (2008) and

Killer Commodities (2008) (where I wrote a chapter on sunscreen). Others include Janice Harper whose critique of development in Madagascar and asthma in Houston continues to develop the political ecology of heath model. Elizabeth Guillette (1998) found that children from two similar towns nestled in the Yaqui Valley, one of Mexico's largest agricultural areas, demonstrated strikingly different neurological capabilities as a result of differential exposure to pesticides. One town was beholden to pesticides since the 1950s while the other town opted for traditional farming and shunned pesticides. Children exposed to pesticides lacked energy, were saddled with significant learning disabilities, and had coordination problems.

Similarly, environmental anthropologists need to work more closely with epidemiologists, as I have with Devra Lee Davis. It is estimated that 10 million cancers over the last 30 years were entirely preventable, most having environmental etiologies as contributing factors. Davis today is bringing widespread attention to the dangers of cell phone use (Davis, 2010).

TROUBLEMAKING ANTHROPOLOGY

Barbara Johnston argues that applied anthropologists can play four different roles: analysts, advocates, mediators, and troublemakers (Johnston, 2001). She warns that "[W]hen environmental justice work involves advocacy and action – confrontational politics – a number of professional bridges are burned. ... 'Cause-oriented' anthropology suggests people who make trouble. Troublemakers are celebrated in this discipline when their cause succeeds and justice prevails. But often 'justice' is elusive, success is hard to gauge, and action results in unforeseen adverse consequences" (Johnston, 2001, p. 8).

Johnston has never held a full time academic job and that is perhaps one reason she has been effective as an activist/troublemaking anthropologist (Johnston, 2011, 2001). Johnston's recent edited work *Half-Lives & Half-Truths, Confronting the Radioactive Legacies of the Cold War* (2007) details the ethnographic (and often advocacy) work of 14 anthropologists researching the consequences of the first nuclear age, from the Marshall Islands to the former Soviet Union. As Laura Nader, a contributor, notes, "over the past fifty-plus years, relatively few American anthropologists or the American Anthropology Association have voiced opposition to this [Marshall Islands] destruction" (p. 304). Nader critiques the scores of anthropologists who worked in the Pacific, like Margaret Mead, but never spoke up about the nuclear testing. She calls this a normative blindness of the profession. At the same time Mead is very important for how she popularized anthropology and worked as a border-crossing intellectual.

Much of the profession is only now learning about historical practitioners of this important work. Earle Reynolds was a physical anthropologist who moved his family to Hiroshima to work as a biostatistician for the Atomic Bomb Casualty Commission's Pediatrics Department in 1951. The more he learned about the devastation and the physical and psychological horrors endured by the Japanese, the more he became convinced that the nuclear arms race had to be stopped. He understood that his work was to survey the bodies of Hiroshimans to predict casualties in a later nuclear war. He quit and became a world renowned activist, sailing his boat into nuclear testing areas to draw attention to the issue. He received letters of support from around the world, even from Martin Luther King Jr., but his colleagues in the American Anthropology Association came to regard him as "a dangerous outsider". Anthropologist David Price used FOIA requests to learn to what extent Reynolds' actions were monitored by the state. He was closely watched. Moreover, as Price notes, "Reynold's activist-applied anthropology radically unhitched mission from employment. His story illustrates how applied research can unleash knowledge that bears a weight of uncomfortable responsibility – a responsibility that if acted upon can lead anthropologists to take duty bound actions ... [that endanger] their careers. That it seems unusual to consider the outlaw Reynolds an applied anthropologist tells us more about the ethical banality we encounter in our work in mainstream applied anthropology than it does about Reynolds".

Yet critical education is the basis of anthropology even though civic minded anthropologists can find themselves at risk simply for doing what they were hired to do: investigate a given external issue. Ted Downing, former SFAA President (1985–1987), experienced this and more. In 1995, Downing wrote an evaluation report describing the severe social and environmental impacts likely to be suffered by Chile's Pehuenche Indians from a proposed dam project underwritten by the World Bank. After his report was censored Downing demanded that the World Bank publicly disclose his findings. The Bank responded by threatening "a lawsuit garnering Downing's assets, income and future salary if he disclosed the contents, findings and recommendations of his independent evaluation." We do not have a good accounting of how often this happens to anthropologists, but we need to learn more about this and teach these lessons to our students. In any case, resisting censorship is, as Downing says, "good applied" anthropology.

Environmental anthropology's future depends on "good applied anthropology." Efforts must be made to dramatically expand the work of these key issues in the discipline: (1) bridging the divide between environmental/ecological and medical anthropology, (2) restoring the holistic connections between the four fields, (3) exploring the phenomenological and religious meanings of environmental and ecological landscapes, (4) expanding the theoretical power of the political ecology of health, (5) learning from archaeologists who are holistic by default, (6) reflexively critiquing the field of anthropology so as to better recognize how state repression continues to limit inquiry, (7) educating academic anthropologists about the perils and possibilities of applied environmental work, (8) overcoming self-censorship and public timidity, (9) adopting theories and practices from ecological socialism, and (10) developing new strategies of applied/activist anthropology centered around journalism, sustainable campus initiatives, critical public pedagogy, and others, yet unnamed.

ON THE DOW DOLE

> Never doubt that a small group of thoughtful, committed citizens can change the world. Indeed, it is the only thing that ever has. (Margaret Mead)

Many universities boast area studies programs that critically investigate the political ecology and culture of specific regions of the world, like Africa, Latin America, or Asia. It's common for these programs to house perspectives that are critical of capitalism. But usually the only sector of the university that studies corporations in an in-depth manner is the Business College, though that's rarely critical. University scholars are not taking full advantage to investigate what Ralph Nader calls "the most animistic entity known to man," the corporations. They are treated like persons! Consider Dow Chemical. It is the 42nd richest company in the world. With revenues of $57.5 billion in 2008, Dow Chemical is worth more than 65% of the world's countries (124 nations), according to World Bank statistics. It's like having a foreign country in your own backyard! Were that Dow could be studied like a foreign country, like it deserves.

I began writing about Dow Chemical after I made a disturbing connection on Michigan State University's campus one day. Dow is a major corporate funder of MSU. In 2001, it merged with Union Carbide and began to receive international rebuke for its refusal to assist the victims of Union Carbide's 1984 Bhopal disaster, the worst industrial accident of all time. Interestingly, a few months later, in the Spring of 2002 Dow co-sponsored a seminar series at MSU's Detroit College of Law, called, "Creating Sustainable Cities in the 21st Century." On March 19th the talk was titled, "Abandonment of the Cities." I noted to myself that there was no mention that day of the irony that Dow Chemical had abandoned the city of Bhopal. Moreover, there were no protests even though MSU had an active sustainability program. A few days later, I wrote about it as a local journalist did (I was then an adjunct at MSU) and then wrote a number of stories about them (McKenna, 2002a, 2002b, 2003a, 2003b, 2004, 2005, 2009).

I propose three areas for publically engaged anthropologists in their work to "green the academy." First, as critical scholars they can lead interdisciplinary team efforts to investigate local hometown corporations in fine ethnographic detail. Second, they can lead their students to research and write a critical environmental histories of their own town or city, carefully documenting how seven local actors (media, medicine, corporations, public health, community groups/unions, environmental groups, and universities) interact to enhance or disturb health. Third, they can join campus sustainability groups (and reach out to serve on Ph.D. committees outside of their departments) and steer them more forcefully toward critical anthropological knowledge and eco-socialist models as they work to ethnographically detail the culture, resources, and power dynamics of the university in reclaiming it as a democratic public sphere.

If one looks squarely at reality they would have to concur with Ralph Nader who famously asserted at an American Anthropological Association conference in 2001 that "anthropologists need not go to all Four Corners of the globe to deconstruct the ideologies of corporate power." For example, he noted, "General Motors has more rights than most U.S. citizens. The most animistic, inorganic institution in the world is the modern corporation ... it's given the constitutional right to remain silent. He said that "studying up means getting behind those images."

The two contradictions of capitalism are on full display. In "greening the academy," anthropologists need to educate citizens about this holistic, cross-cultural, and necessarily radical (or root) view of reality. As critical intellectuals, anthropologists are in a position to forcefully critique – in public – their own cities, hometowns (and universities), diagnosing how issues of culture, resources, and power impact local health and environment. Anthropologists need to bring the full weight of anthropological knowledge, insight, and methodologies to wider audiences. And they need to rethink how to cast these activities as a continuation of the sentiments expressed in the unfinished American Revolution. The green movement provides an important space for anthropologists to engage in Thomas Paine's "revolutions of justice." As CUNY anthropology professor David Harvey states it, "The capitalist class will never willingly surrender its power. It will have to be dispossessed" (Harvey, 2010).

ACKNOWLEDGMENTS

I would like to thank several anthropologists for generously contributing their time to reviewing this chapter and making valuable contributions: Les Sponsel, Barbara Johnston, Colleen Boyd, Peggy Barlett, Carl Maida, Sam Beck, Gene Anderson and Pamela Puntenney. All errors are my own.

REFERENCES

American Anthropological Assn. (2011). *Fieldschools*. Retrieved from www.aaanet.org/profdev/fieldschools/

Anderson, E. N. (1996). *Ecologies of the heart: Emotion, belief and the environment*. New York, NY: Oxford University Press.

Aronowitz, S. (2000). *The knowledge factory: Dismantling the corporate university and creating true higher learning*. New York, NY: Beacon.

Baer, H., & Singer, M. (Eds.). (2008). *Killer commodities: Public health and the corporate production of harm*. New York, NY: AltaMira.

Balee, W. (1996). Anthropology. In Collett & Karakashian (Eds.), *Greening the college curriculum. A guide to environmental teaching in the liberal arts*. Washington, DC: Island Press.

Barlett, P., & Chase, G. (Eds.). (2004). *Sustainability on campus: Stories and strategies for change*. Cambridge, MA: MIT Press.

Bateson, G. (1972). *Steps to an ecology of mind*. Chicago, IL: University of Chicago Press.

Bauman, Z. (2006). *Liquid fear*. Cambridge: Polity.

Beck, S. (2009). Introduction: Public anthropology. *Anthropology in Action, 16*(3), 1–4.

Berlin, E. A., & Berlin, B. (1996). *Medical ethnobiology of the Highland Maya of Chiapas, Mexico: The gastrointestinal diseases*. Princeton, NJ: Princeton University Press.

Biolsi, T. (Ed.). (2004). *A companion to the anthropology of American Indians*. Malden, MA: Blackwell.

Brockington, D. (2011). *Environmentalism and conservation*. Retrieved from http://environmentalismandconservation.wordpress.com/

Buscher, B. (2011). *The critical side to conservation and energy*. Retrieved from http://brambuscher.com/

Carrier, J. G., & West, P. (2009). Virtualism, governance and practice: Vision and execution in environmental conservation. In *Studies in environmental anthropology and ethnobiology*. New York, NY: Berghahn.

Casagrande, D. G., Hope, D., Farley-Metzger, E., Cook, W., & Yabiku, S. (2007). Problem and opportunity: Integrating anthropology, ecology, and policy through adaptive experimentation in the urban American Southwest. *Human Organization, 66*(2), 125–139.

Collins, J., di Leonardo, M., & Williams, B. (2008). *New landscapes of inequality, neoliberalism and the erosion of democracy in America*. Santa Fe, NM: SAR Press.

Davis, D. (2000). *The secret history of the war on cancer*. New York, NY: Basic.

Denevan, W. M. (1996). Pristine myth. In D. Levinson & M. Ember (Eds.), *Encyclopedia of cultural anthropology* (Vol. 3, 1034–1036). New York, NY: Henry Holt and Co.

Dove, M., & Carpenter, C. (Eds.). (2005). *Environmental anthropology: A historical reader*. Malden, MA: Blackwell.

Dressler, W. (2011). *Politics, environment and development*. Retrieved from http://wolframdressler.com/

Duffy, R. (2011). *Conservation politics*. Retrieved from http://conservationpolitics.wordpress.com/

Eriksen, T. H. (2006). *Engaging anthropology: The case for a public presence*. London: Bloomsbury.

Ervin, A. (2005). *Applied anthropology: Tools and perspectives for contemporary practice* (2nd ed.). Boston, MA: Allyn and Bacon.

Fletcher, R. (2011). *Cultural dimensions of ecotourism and conservation*. Retrieved from http://ecotourismconservation.wordpress.com/

Freire, P. (1970). *Pedagogy of the oppressed*. New York, NY: Seabury Press.

Giroux, H. (2004). *The terror of neoliberalism*. Boulder, CO: Paradigm Publishers.

Giroux, H. (2007). *The university in chains: Confronting the military-industrial-academic complex*. Boulder, CO: Paradigm Publishers.

Giroux, H. (2010). *Zombie politics and culture in the age of casino capitalism*. (Popular Culture and Everyday Life.) New York, NY: Peter Lang Publishing.

Giroux, H. (2011a, May 16). American democracy beyond casino capitalism and the torture state. *Truthout, 23*(3), 2–6.

Giroux, H. (2011b). *The twilight of the social: Resurgent politics in an age of disposability?* Boulder, CO: Paradigm Publishers.

Guillette, E. (1998). An anthropological approach to the evaluation of preschool children exposed to pesticides in Mexico. *Environmental Health Perspectives, 106*, 347–353.

Harris, M. (2001 [1969]). *The rise of anthropological theory. A history of theories of culture*. Walnut Creek, CA: Alta Mira.

Harris, M. (2001 [1979]). *Cultural materialism: The struggle for a science of culture*. Walnut Creek, CA: Alta Mira.

Harvey, D. (2009). *Cosmopolitanism and the geographies of freedom*. New York, NY: Columbia University Press.

Harvey, D. (2010). *The enigma of capital: And the crises of capitalism*. Oxford: Oxford University Press.

Hornborg, A., & Crumley, C. L. (Eds.). (2006). *The world system and the Earth system: Global socioenvironmental change and sustainability since the Neolithic*. Walnut Creek, CA: Left Coast Press.

Igoe, J. (2011). *Spectacle of nature*. Retrieved from http://spectacleofnature.wordpress.com/

Ingold, T. (2000). *The perception of the environment: Essays on livelihood, dwelling and ill*. London: Routledge.

Johnston, B. (1994). *Who pays the price? The sociocultural context of environmental crises*. Washington, DC: Island Press.

Johnston, B. (2011). *Life and death matters: Human rights, environment and social justice* (2nd ed.). Walnut Creek, CA: Left Coast Press.

Johnston, B. R. (2001). Anthropology and environmental justice: Analysts, advocates, mediators, and troublemakers. In C. Crumley (Ed.), *New directions in anthropology and environment: Intersection*. Lanham, MD: AltaMira.

Johnston, B. R. (Ed.). (2007). *Half-lives & half-truths: Confronting the radioactive legacies of the cold war*. Santa Fe, NM: School for Advanced Research Press.

Jones, A. (2010). Michael Halliday: An appreciation. *International House Journal of Education and Development*, (28), 13–16.

Kahn, J. (2003). Making Iraq safe for capitalism: can a U.S. Treasury team build a free-market economy in a war-torn land? *Fortune*, July 7.

Lansing, S. (1991). *Priests and programming: Technologies of power in the engineered landscape of Bali*. Princeton, NJ: Princeton University Press.

Maida, C. (Ed.). (2007). *Sustainability and communities of place*. New York, NY: Berghahn.

Mandel, M. (2005). *Nuremberg lesson for Iraq War: It's murder*. Retrieved from Commondreams.com. Accessed on August 30.

Mann, C. (2005). *1491: New revelations of the Americas before Columbus*. New York, NY: Knoph.

Mattei, U., & Nader, L. (2008). *Plunder, when the rule of law is illegal*. Malden, MA: Blackwell.

Mayberry-Lewis, D. (1992). *Millennium: Tribal wisdom and the modern world*. New York: Penguin.

McElroy, A., & Townsend, P. K. (2009). *Medical anthropology in ecological perspective* (5th ed.). Philadelphia, PA: Westview.

McKenna, B. (2002a, December–January). Environmental data suppression in Lansing: Why did they do it? *From the ground up* (published bi-monthly by the Ecology Center, Ann Arbor, MI), *33*(4), 15–17.

McKenna, B. (2002b). The Hollister/GM offensive: Fear and profits versus the peoples' health. *The City Pulse*, May 1, pp. 1, 6, 7.

McKenna, B. (2003a). Who is Michigan's empire man? Big Ten University president does Bush's bidding. *Free Press*, August 18.

McKenna, B. (2003b). Education for what? A chronicle of environmental health deception in Lansing, Michigan. *Cooley Law Review*, *20*(2), 1–54.

McKenna, B. (2004). On the Dow Dole, can Michigan's universities be critical of chemical giant? *From the Ground Up* (pp. 1, 7–16) (published bi-monthly by the Ecology Center, Ann Arbor, MI). Cover Story, January.

McKenna, B. (2005). Dow Chemical buys silence in Michigan, documentary on Dow's Dioxin scandal ignored by four local PBS stations. CounterPunch. Petrolia, CA, April 18.

McKenna, B. (2006). Native resurgence spurs home, giving thanks to America's Indians. CounterPunch, November 24.

McKenna, B. (2008a, August). The anthropology of censorship: "Yes, But ...". The Heggenhougen Challenge. *Society for Applied Anthropology Newsletter*, *19*(3), 15–19.

McKenna, B. (2008b). Higher ed's 'civic engagements' dumbed down: Conjuring Freire in Dearborn. CounterPunch.

McKenna, B. (2009a). Dow Chemical's knowledge factories: Action anthropology in Michigan's company town culture. *Anthropology in Action, Journal for Applied Anthropology in Policy and Practice*, *16*(2), 39–50.

McKenna, B. (2009b). Pigs for the ancestors, American style. *Society for Applied Anthropology Newsletter*, *20*(3), 16–20.

McKenna, B. (2011a). Bestselling anthropologist 'Predicted' the financial meltdown of 2008. *Society for Applied Anthropology Newsletter*, *22*(1), 21–25.

McKenna, B. (2011b, December). Staging a Christopher Columbus play in a culture of illusion: Public pedagogy in a theatre of genocide. In B. McKenna & A. Darder (Eds.), *Policy futures in education* (Vol. 9, Issue 6, pp. 670–685).

Merchant, C. (2005). *Radical ecology: The search for a livable world.*

Moran, E. (2010). *Environmental social science: Human environment interactions and sustainability*. Oxford: Wiley/Blackwell.

Mühlhäusler, P., & Peace, A. P. (2006). *Annual Review of Anthropology*, *35*, 457–479.

Nazarea, V. (2006). *Cultural memory and biodiversity*. Tucson, AZ: University of Arizona Press.

O'Conner, J. (1998). *Natural causes, essays in ecological Marxism*. New York, NY: Guilford.

Oswalt, W. H. (2009). *This land was theirs, A study of native North Americans*. London: Oxford.

Pauketat, T. (2009). *Cahokia: Ancient America's great city on the Mississippi*. New York, NY: Viking.

Price, D. (2004). *Threatening anthropology McCarthyism and the FBI's surveillance of activist anthropologists*. Durham, NC: Duke University Press.

Public Employees for Environmental Responsibility (PEER) (2001). *The story of water resources at work, Ingham County, Michigan*, Washington, DC, 130 pages, September 19.

Rappaport, R. (1968). *Pigs for the ancestors: Ritual in the ecology of a New Guinea people*. New Haven, CT: Yale University Press.

Redman, C. (1991). *Perspectives on Southwestern prehistory*. Denver, CO: Westview.

Robbins, R. (2008). *Cultural anthropology: A problem-based approach*. Belmont, CA: Wadsworth.

Sahlins, M. (1972). *Stone age economics*. Piscataway, NJ: Aldine Transaction.

Sandlin, J., Schultz, B., & Burdick, J. (Eds.). (2009). *Handbook of public pedagogy: Education and learning beyond schooling*. New York, NY: Routledge.

Singer, M. (2010). Eco-nomics: Are the planet-unfriendly features of capitalism barriers to sustainability? *Sustainability*, *2*, 127–144.

Steward, J. (1955). *Theory of culture change: The methodology of multilinear evolution*. Urbana, IL: University of Illinois Press.

Sullivan, S. (2010). Green capitalism, and the cultural poverty of constructing nature as a service provider. *Radical Anthropology*, 18–27.

Sullivan, S. (2011). Engagements with nature, culture and capitalism. Retrieved from http://siansullivan.wordpress.com/about/

Tett, G. (2009). *Fool's gold: The inside story of J.P. Morgan and how Wall St. Greed corrupted its bold dream and created a financial catastrophe*. New York, NY: Free Press.

Wallis, V. (2010). Beyond green capitalism. *Monthly Review*, *61*, 9. Retrieved from http://monthlyreview.org/2010/02/01/beyond-green-capitalism

Wells, E. C. (2010). Anthropology, sustainability, and higher education: An interview with Peggy Barlett. *Society for Applied Anthropology Newsletter*, *21*(4), 18–22.

West, P. (2006). *Conservation is our government now: The politics of ecology in Papua New Guinea (New ecologies for the twenty-first century)*. Durham, NC: Duke. Retrieved from http://bc.barnard.edu/~pwest/index.html

THE FERGUSON SHOOTING, 2014: A SPATIAL AND MEDIA ANALYSIS

Liam Leonard

ABSTRACT

This chapter will report on the issues that surround the coverage of police shootings in the United States. It will also look at the issue of race and policing. In particular, the events at Ferguson, St. Louis in August 2014 and other controversial incidents will be explored. The chapter will utilize theories of policing, employ a media analysis, and develop an issue history to explain the background to problems with policing in African-American neighborhoods over many decades, culminating in the rise of the Black Lives Matter movement.

Keywords: Ferguson; police shootings; spatial analysis; media analysis; criminology; urban

INTRODUCTION

This chapter provides a media analysis of the response to events surrounding the shooting of Michael Brown in Ferguson, St. Louis in August 2014. The goal of the chapter is not to ascertain the guilt of the parties involved in this incident. It does however attempt to present an analysis of the editorial framing agendas of certain media groups which covered the incident.

These media groups were chosen for their online content and editorial policies which reflected the framing of events and people involved in the

Environmental Criminology: Spatial Analysis and Regional Issues
Advances in Sustainability and Environmental Justice, Volume 20, 43–56

ISSN: 2051-5030/doi:10.1108/S2051-503020170000020003

Ferguson case. Online media outlets were also utilized, from the evening of the Brown shooting in August 2014 through to the official response of President Barak Obama to the shooting, protests and Grand Jury announcement in November and December 2014.

The chapter attempts to provide an understanding of media biases during this coverage, in order to add to the debates on key issues such as media, policing, and race in this case and in the future. Key events from August 2014 through to January 2015 are covered.

Both spatial and media analysis are utilized in this chapter. Spatial Analysis is used to identify the problems facing the inhabitants of Ferguson, an African-American section of St. Louis in the United States. Media analysis can take on many forms. In this case, the slant of each media group is reported, and categorized as radical, liberal, mainstream, or conservative.

ASSESSING MEDIA VALUATION STRATEGIES

This chapter also employs a media analysis developed through an *evaluation strategy*. "Evaluation strategies" are broad, overarching perspectives on evaluation. The evaluation strategy utilized is the qualitative/anthropological model which emphasizes the importance of observation, the need to retain the quality, and ideology of evaluation process. Media theory presents us with an understanding of mass media. In recent years, this has been extended into the area of social media through the rise of the internet derived "Network Society" (Castells, 2000).

By applying a Qualitative Content Analysis of the Evaluation Strategies behind media coverage of the events at Ferguson, this study will provide an evaluation of how mainstream and social media reported on what occurred in the St. Louis suburb on the night of Michael Brown's death, and in the weeks subsequent to that event.

Meanings will be elucidated from the media event analysis, which will provide findings and establish understandings of the relevant issues surrounding the issues of criminal justice, media, and race both in St. Louis and across the United States.

Media analysis of the shooting of Michael Brown on August 9, 2014, an unarmed 18-year-old youth was shot and killed by a police officer in Ferguson, Missouri, a suburb of St. Louis. The youth, Michael Brown, was an African-American, while the officer involved in the shooting, Darren Wilson, was white. In the subsequent days and weeks which followed, crowds of protests would take to the streets of Ferguson to protest about the shooting. The police response to these protests caused even further concern for many Americans watching news reports on television and online. These concerns were heightened by the appearance of militarized armored vehicles and police dressed in military style body armor. For a period of time, the streets of Ferguson began to resemble a besieged city in Iraq or Afghanistan, as the police response took on the appearance of a military occupation.

Many US citizens were shocked to see such images in the American heartland. Policing in the United States had seen to reach a disturbing turning point. Another shocking aspect of some of the mainstream coverage was the hostile depiction of African-Americans in the story, from street protestors to family members and advocates.

The report will provide coding frames to assist with the creation of pattern analysis of contributing events in relation to the media's coverage of the criminal justice system during these events. This framework is part of the "Issue History" (Szasz, 1998) which can be developed around such key factors.

This case study argues that the events in Ferguson were part of a longstanding history of controversy in the areas of race and criminal justice in the United States, since the 1800s. This "Issue History" can be seen in the key events and outcomes during this time.

The shooting of Michael Brown by police on August 9, 2014 in Ferguson was covered by many mainstream media outlets. NBC News featured the Brown shooting by incorporating nearly 300 storylines about events at Ferguson onto its online edition. Therefore, a media analysis based on the coverage of NBC News, as a national media outlet, will provide the first part of this wider investigation.

The first NBC report on the events which occurred at Ferguson is dated August 10, 2014. The dichotomous approach to this story is immediately indicated by the controversial headline "Kill the Cops" which is carried over a photo of local protestors holding their hands up as part of what would be known as the "Don't Shoot" stance taken by participants who wanted to replicate the last stance and words of Michael Brown, as reported by witnesses. The "Don't Shoot" hands up stance would become a central image of the Ferguson protests.

In the report, Ferguson is described as "a predominantly black suburb north of downtown St. Louis." After describing how the shooting victim's grandmother found the body in the street near his home, the police response provided was non-committal:

> A spokesman with the St. Louis County Police Department, which is investigating the shooting at the request of the local department, confirmed a Ferguson police officer shot the man but didn't give the reason for the shooting. (nbcnews.com. August 10, 2014)

The report then outlines the response to the incident, stating a crowd gathered outside the Ferguson police station shouting slogans such as "Kill the police." The article ends with a photo of Lewis Head, Michael Brown's stepfather, who holds a sign saying, "Ferguson Police just executed my unarmed son." The other main photo is of the crowd confronting police officers who are standing behind a yellow crime scene tape.

Both images establish two of the main narratives of the Ferguson case; one is of the personal tragedy to the Brown family while the other highlights the emotional response of the wider African-American community in the area. The coding of the personal and the public aspects of this issue will be among the issues covered in this study.

The initial coverage by nbcnews.com of both aspects sets the tone for the overall coverage of what occurred at Ferguson, and both require scrutiny to provide a fuller understanding of the key issues involved.

This "public/private" approach is often utilized by media outlets for political coverage such as electoral campaigns. The candidate's public statements are contrasted with private behavior in order to create a moral code filter where moral judgments and condemnations can be applied.

This moral filter gives media its power over political events and allows the media to breakdown events for its client audience, with added values such as ideology then included through editorial control. In the initial nbcnews.com coverage, the public have been given a view of the Brown family as powerless and unconnected (as with the photo of Michael Brown's stepfather holding a torn piece of cardboard with a scrawled inscription), while the public depiction of the Ferguson community was shown to be unruly in the face of a stoic police line. The initial lines of media coverage had been drawn, and all subsequent coverage would follow along these lines.

As the discipline of Media Studies provides us with theories and analysis of today's mass media, we are able to decipher some of the coda underpinning media coverage of volatile events such as those which occurred at Ferguson. McLuhan (1994) provided an understanding that "the media is the message" and that this shapes our understanding of the world and of the key events which shape it.

For McLuhan, the media and its leaders, gatekeepers and consumers in the "global village" collectively form a "media ecosystem" where messages are presented, interpreted and reinterpreted, constantly creating new meanings and changing understandings of key events as part of a social discourse.

ESTABLISHING MEDIA HEGEMONY

The second report dated August 10 in this analysis combines footage from the police conference with a report on events. This report outlines the police department's version of events surrounding the shooting:

> St. Louis County Police Chief Jon Belmar said Michael Brown, of Ferguson, Missouri, was struck "more than just a couple" times by gunshots fired by a Ferguson Police Department officer after a "physical confrontation" that began in a police vehicle and then spilled on to the street. The officer is now on paid administrative leave, Belmar said. The Ferguson Police Department reached out to Belmar immediately following the incident requesting that the St. Louis County Police Department launch an independent investigation, he said. (nbcnews.com August 10, 2014)

The police department press conference was reported as a definitive version of events, one which through doubts on eyewitness claims that the victim was shot while surrendering. In addition, footage of protestors provided a demarcation between the police reports and the response of community; in so doing, the media can be seen to be building a framework which favored the authorities and cast doubt on the witnesses' accounts, thereby casting the unruly protestors as responding in a visceral and ultimately lawless fashion.

A further dichotomy is established in the report as police are shown to "reach out" to Chief Belmar in St. Louis.

Conversely, the National Association for the Advancement of Colored People (NAACP) is presented outside of this local "reaching out" as they called for a wider response from the justice sector:

> The shooting has sparked an outcry in the predominantly black town of about 21,000 – a suburb of the city of St. Louis – and the local chapter of the NAACP have launched their own investigation into the case, appealing to the FBI for help. "We are hurt to hear that yet another teenage boy has been slaughtered by law enforcement," St. Louis County NAACP President Esther Haywood said in a statement. (nbcnews.com, August 10, 2014)

The Ferguson community is depicted as an outside community to the rest of the city in these reports ("the predominately-black town of about 21,000") while the police are depicted as requiring outside support from the start. Implicitly, the reports are establishing the Ferguson community as a potential threat, to the police and to the security of the St. Louis region. This in some way provides congruence for the police report which indicated "a struggle" before Michael Brown was shot and killed, again running against witness statements.

The end result of this form of reporting is one which establishes an understanding that the authorities are telling the truth while the community (victim's family, witnesses, community spokespersons, and protestors) must therefore be presenting a version of events which are in some way untrue, rendering any further responses from this group to be somewhat outside the truth and ultimately, outside the law. This is a quintessentially "hegemonic" response from the media, as outlined by the theorist Antonio Gramsci (1982).

Hegemony is the use of language in the establishment of relations of power between those in the establishment and the rest of society. In contemporary society, it is cultural institutions such as the mainstream media which provide the platform for the establishment of hegemony, as television and newspapers set out understandings of social and political events according to the agendas of those in power. While the reports on Ferguson may have done so unintentionally, the modern media industry provides a cultural aspect to the establishment, as a part of what the writer Noam Chomsky (2002) has described as "Manufacturing Consent."

MEDIA LABELING OF PROTEST

The third group of reports focused on the community's response to the shooting. Headlines varied from "Looting Erupts After Vigil for Slain Missouri Teen Michael Brown" to "Missouri Teen's Fatal Shooting by Cops Sparks Riots," and "Looters Raid Store After Cop Shoots Teen" (NBC News, August 11, 2014). Accompanying photos show looters ransacking a grocery store.

While looting has often taken place after public protests, there was no attempt made in the reports to differentiate between the protests which emerged spontaneously after Michael Brown's shooting and the opportunistic looting which followed. No one involved in the protests had advocated looting, although anti-police sentiment was openly expressed.

Essentially, the reports had run a thread between the response of the Brown family and Ferguson community which had mobilized in a show of support for them after the shooting, and the criminal behavior of opportunistic looters who had taken advantage of the turmoil in the area. To the outside observer, a large section of the community at Ferguson had been criminalized, despite the fact that no one had at that point been made accountable for the shooting of an unarmed youth in the neighborhood.

Moreover, while there was no direct evidence of a link between the protestors and the looting which came afterwards, the report drew a thread between both events and their participants:

> A day of anger over a fatal police shooting of an unarmed black man in suburban St. Louis turned to mayhem as people looted businesses, vandalized vehicles and confronted police... The tensions erupted after a candlelight vigil...Afterward, a convenience store was looted. Several other stores along a main road near the shooting scene were broken into, including a check-cashing store, a boutique and a small grocery store. (nbcnews.com, August 11, 2014)

The reports on looting were interspersed with coverage of the protests. While police chiefs appealed for calm, the FBI were said to be investigating the incident. The family of Michael Brown released a statement saying: "We need Justice." As protests continued, police used tear gas against participants. One issue which emerged was the refusal of police to disclose the name of the officer involved in the shooting. This led to suspicions of a cover up.

By August 18, the Missouri Governor Jay Nixon had called a State of Emergency and issued a curfew in Ferguson. Police had released a video which appeared to show Michael Brown stealing a box of cigars from a local store, while pushing a store owner aside in the process. The release of this video enraged protestors further, who saw it as an attempt to smear Brown's name in the aftermath of his shooting.

Mainstream media reports stated that "armored vehicles rolled back onto the streets of Ferguson ... as riot police faced off with looters." The blurring of lines between "protestors," "rioters," and "looters" continued, even when national figures joined the protests:

> Protests had started off peacefully in Ferguson on Friday night. Rev. Jesse Jackson linked arms with protesters, leading them in prayer and urging them to "turn pain into power" while fighting back non-violently. Shortly after midnight, crowds got rowdier and looting began to break out. (nbcnews.com, August 16, 2014)

The national media contained many images of a militarized police force being deployed at Ferguson in a manner which surprised many observers. As protestors adopted the "Hands Up-Don't Shoot" surrender pose which witnesses said Michael Brown took just before his shooting, the police lines which faced them included militarized vehicles and weaponry. Radley Balko's *Rise of the Warrior Cop* (2014 Public Affairs: New York) deals with the onset of this form of militarized policing. The "Hands Up" slogan became central to the Black Lives Matter movement which grew from this and other similar shootings.

This militarized equipment was more readily associated with battles during the Iraq War than with policing in the American heartland. Furthermore, while many accepted the need for a strong response to looting, the deployment of such a militarized force against Americans in their own neighborhoods was surprising and disturbing to many. By August 19, after 10 days of protests, the Missouri National Guard had been deployed. Attorney General Eric Helder stated he would travel to Ferguson to try to intervene. This wider response came as the protests continued, and the militarized style policing response intensified. The release of the autopsy on Michael Brown showed the extent of his injuries, angering protesters further.

SPATIAL ANALYSIS OF TWO SHOOTINGS: LONDON AND FERGUSON

One aspect of the police response was the arrest of observers, journalists, and others in a heavy-handed crackdown in the Ferguson area. A State Senator was arrested at a protest while two journalists were arrested at a local restaurant for not evacuating quickly enough:

> State Senator Jamilah Nasheed, a Democrat from St. Louis, was arrested outside the police station for walking and standing in the street in violation of a city ordinance, according to St. Louis County police, who are helping patrol the streets. (nbcnews.com, August 21, 2014)

Respected Princeton academic and rights advocate Cornel West was arrested along with 49 other protestors on the street and at a local Walmart. Professor West spoke at an Interfaith Rally on October 12, which drew an attendance of 2,000 people. The meeting also featured contributions from ordinary protestors who voiced their objection to the Brown shooting. It is also interesting to examine the coverage given to external participants by certain media groups. *The Guardian Online* presented coverage of the arrest of two journalists early in the protests on August 13 and 14, 2014. The Guardian outlined the arrest of the journalists as follows:

> Two reporters were arrested and teargas was fired at demonstrators in Ferguson, Missouri, on Wednesday, five nights after an unarmed black teenager was shot dead by an officer from the town's predominantly white police force. The Washington Post condemned the detention of its journalist, Wesley Lowery, as "illegal" and an "assault on the freedom of the press to

> cover the news." The Huffington Post attacked the arrest of its reporter, Ryan J. Reilly, as "militant aggression." Reilly said that the "police resembled soldiers more than officers." (The Guardian Online, August 13, 2014)

It is worth noting the different slant taken by the traditionally liberal leaning *Guardian*, (a US edition of a British newspaper) in comparison with the more conservative *NBC Online* coverage discussed earlier. Here, the Guardian seems to have captured the prevailing mood of both Americans who were shocked at the militarized appearance of the police deployed at Ferguson and the concerns expressed by both the *Washington Post* and the *Huffington Post* at the treatment of their reporters at the hands of this police force.

It is also interesting to note that the Guardian stresses the fact that Michael Brown was "an unarmed teen" and that Ferguson's police were "predominately white." This coverage is very different to *NBC Online's* depiction of unruly protestors and out of control looters.

The Guardian's Jon Swaine witnessed the arrest of two reporters, and described it as follows:

> The two reporters were taken in cuffs from a McDonald's in downtown Ferguson and placed in an unmarked white police van. "They're reporters," shouted one member of the public, who had also been in the restaurant. When I approached a police line in an effort to confirm the reporters' identities, I was threatened with arrest and told to retreat. (Guardian Online, August 13, 2014)

The Guardian can also be distinguished from the coverage of *NBC Online* due to its utilization of academic perspectives of the events at Ferguson. The British criminologist Tim Newburn provides a discussion of how Ferguson differs from previous British race related protests.[1] In his article, Newburn contrasts the 2011 shooting by British police of Mark Duggan, and unarmed black teenager from London, with that of Michael Brown in Ferguson. Newburn cites a report commissioned by the London School of Economics and *the Guardian* called "Reading the Riots" that provided the following findings:

> Beyond the shooting itself, significant alienation among young people and deep distrust of, and hostility towards, the police were key factors in the rioting in all the cities in which disorder occurred. (Guardian Online, August 20, 2014)

Despite the similarities in both cases, where young unarmed black men were shot by police, followed by a period of unrest, Newburn outlines the differences in the UK and US incidents. What becomes crucial from Newburn's perspective was the outbreak of "an information war" in both cases, almost from the moment the shootings took place. And while policing in the United States and United Kingdom is very different in relation to arms, the two shootings have similarities.

Initially, the police released details, including stills from a surveillance camera, of a convenience store robbery that had occurred just before the shooting, suggesting that Brown had been involved. Details of the initial autopsy on Michael Brown were then made public, including the suggestion that he had traces of cannabis in his bloodstream.

As family and friends had felt in the Duggan case, it appeared that strenuous efforts were being taken to divert attention from the shooting itself. The rioting, though it has not as yet spread beyond Ferguson, has exhibited many of the elements that were evident in England in 2011: attacks on police, the looting of local stores, the widespread destruction of property, and complaints about police harassment (The Guardian, August 20, 2014).

Nonetheless, Newburn outlines some distinctive aspects to both events. While he sees Ferguson as being very much about race, it was only partially so in the United Kingdom after Duggan was shot. (Duggan was part Afro-Caribbean and part Irish, both communities which have suffered at the hands of the UK police.) In the aftermath of a series of shootings of black youth in the United States such as Trayvon Martin killing in February 2012,[2] the killing of Michael Brown, another unarmed African-American youth, was seen to be a tipping point in community relations with the authorities across the United States.

Newburn sees real similarities between Ferguson and the UK's largest riots by the Afro-Caribbean community in South London, Birmingham, and Liverpool in the early 1980s. These riots had contributory factors which combined race and policing, all issues which had been instrumental in the events in Ferguson.

The manner in which predominately white police forces had policed black communities in 2011 London and 2014 St. Louis shaped the responses of local communities angered by the killing of unarmed youths from their neighborhoods. Newburn also sees differences in the response of the UK police and the militarized force which descended on Ferguson, and notes that the Missouri Highway Patrol had brief success in calming the situation:

> As illustrated by the brief interlude in Ferguson last week when the highway patrol managed briefly to quell the disorder, a combination of firm but proportionate policing with serious engagement with the community is generally viewed as holding out the greatest promise of success. (Guardian Online, August 20, 2014)

Conservative News Coverage

As this report is covering all forms of media, the coverage afforded to the Ferguson issue by conservative media outlets Fox News has been included. This also adds a right wing perspective to what is largely a survey of Liberal and Radical news and social media outlets. Fox has been criticized for having a right wing, white-centric perspective, with mostly white reporters and presenters. Some conservative presenters and journalists have been seen to be hostile to minorities, and seemingly racist sentiments have been expressed on the channel occasionally. Fox News has also been criticized as a type of "info-tainment" channel rather than a de facto news outlet.[3]

In addition, some conservative contributors made grossly inaccurate claims in the aftermath of the shooting of Michael Brown. These contributors were appeared regularly when other channels were discussing race relations after the incident at Ferguson. For instance, contributor Heather Mac Donald claimed that criminologists were exaggerating injustices in the system:

> The criminology profession has been trying for decades to prove that the overrepresentation of blacks in prison or in arrest statistics is a result of criminal justice racism. It is black crime rates that predict the presence of blacks in the criminal justice system, not some miscarriage of justice. (H. MacDonald, *Meet the Press*, August 17, 2014)

Subsequently, Fox Radio's Todd Starnes accused President Obama and his Administration of "Orchestrating the Michael Brown Tragedy" and claimed the Administration was working with activist leaders such as the Reverend Al Sharpton toward this goal. According to Starnes:

> The uncivil unrest in Ferguson, Missouri continues, and it appears the Obama Administration may be orchestrating the Michael Brown tragedy. NBC News is reporting that Attorney General Eric Holder and Valerie Jarrett are the administration's point persons going so far as to arrange a federal autopsy .... So while President Obama whacks golf balls and dances the two-step in tony Martha's Vineyard, his administration is in cahoots with a bona fide race hustler. (http://mediamatters.org/issues/racial-justice)

Another distortion tactic of right wing commentators was the promotion of a discussion on "black on black" gun crime, which sought to deflect attention from the Ferguson shooting and the issue of inadequate criminal justice responses when African-Americans are involved in shooting incidents at the hands of police departments.

By the end of August, Fox News programs expressed concern that references to Michael Brown as an "Unarmed Teen" might be "misleading" some sections of the public. This came as a smear campaign of sorts began against Michael Brown, with the video of the deceased's involvement in an alleged altercation at a convenience store being released by the authorities.

While Fox presenter Bill O'Reilly launched a campaign which argued against the existence of "White Privilege" and the station ran stories against extending voter registration, other contributors made further accusations surrounding the Brown case, such as Fox contributor Bernie Goldberg's claims that "Facts Mean Something Totally Different" to "Black Liberals" in the aftermath of the shooting of Michael Brown and that "Slavery, Segregation, And The Humiliation That Blacks Suffered Under Jim Crow Has Turned Some People Socially Paranoid" (The O'Reilly Factor, October 23, 2014).

Online Survey Debates

While the mainstream media examined the issue in a traditional manner, the online discussion was sometimes more nuanced. This section will examine media discussions about online surveys on Ferguson. The sites involved are YouGov and the Pew Research Center, with the results being discussed on *Huffington Post*.

According to the Pew Research Survey, there is a marked difference in the responses of black and white people to the Ferguson shooting and aftermath:

> Blacks are about twice as likely as whites to say that the shooting of Michael Brown raises important issues about race that need to be discussed. (Pew Research National Survey, August 14–17, 2014)

The survey also outlined variations in the black and white population in relation to the police response and investigation to the shooting. The survey of 1,000 people, taken in the aftermath of the shooting, demonstrates certain differences in both sets of the population on many of the key questions surrounding the incident. The survey asks whether:

Important issues about race or whether the issue of race is getting more attention than it deserves: 44% think the case does raise important issues about race that require discussion, while 40% say the issue of race is getting more attention than it deserves.

By about four-to-one (80% to 18%), African-Americans say the shooting in Ferguson raises important issues about race that merit discussion. By contrast, whites, by 47% to 37%, say the issue of race is getting more attention than it deserves (Pew Research Survey, August 14–17, 2014).

While many in the African-American population were concerned about the Brown shooting in the aftermath of similar deaths of young African-Americans such as Trayvon Martin, many whites seem to think that too much focus was being placed on race. This understanding was clearly influenced by the media coverage of the Ferguson incident, and serves as a reminder of the need for balanced coverage of incidents involving race and policing.

Pew Research Center Survey August 14–17, 2014.

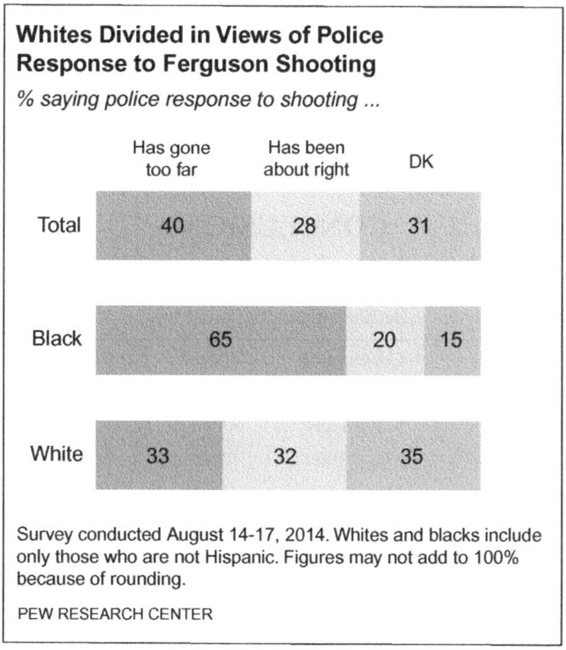

This racial divide can also be seen in relation to the question of whether the police went too far in their response to the Ferguson shooting, though the deploying of a militarized force with heavy armored vehicles for instance:

Fully 65% of African-Americans say the police have gone too far in responding to the shooting's aftermath. Whites are divided: 33% say the police have gone too far, 32% say the police response has been about right, while 35% offer no response.

Three quarters of African-American respondents had little confidence in the judicial process. This can be compared with roughly one-third of whites who responded; this sector felt the justice system was fair and could be trusted. Therefore, the challenge remains to those in the justice system; how can this gulf in confidence be overcome?

In relation to the Brown shooting, the Pew survey found that there was a political as well as racial divide in the responses received. For those who replied from both races, 68% of those who identified themselves as Democratic voters found that the Brown case raises important issues about race, while 61% of those who identified themselves as Republicans felt the Ferguson was over covered.

The finding of this survey reveal a racial and political divide which is centered on the media coverage of events such as the Ferguson shooting and its aftermath. According to Robert Jones in *The Atlantic*:

> the Pew Research Center poll found that overall the country is divided over whether Brown's shooting "raises important issues about race that need to be discussed" (44 percent) or whether "the issue of race is getting more attention than it deserves" (40 percent). However, black Americans favor the former statement by a four-to-one margin (80 percent vs. 18 percent) and at more than twice the level of whites (37 percent); among whites, nearly half (47 percent) believe the issue of race is getting more attention than it deserves. ("Self-Segregation: Why It's So Hard for Whites to Understand Ferguson," The Atlantic, August 14, 2014)

CONCLUSION

	Full Cover	Headline Bias	Editorial Bias	View	Coverage Type
NBC News	Yes	Conservative	Closed	Top down	Neutral
CNN	Yes	Liberal	Open	Middle	Neutral
Washington Post	Yes	Mainstream	Open	Middle	Neutral
New York Times	Yes	Mainstream	Open	Middle	Neutral
Huffington Post	Yes	Liberal	Open	Middle	Pro-protests
AlterNet	No	Radical	Organic	Bottom up	Pro-protests
Daily Kos	No	Radical	Organic	Bottom up	Pro-protests
Fox News	No	Conservative	Closed	Top down	Anti-protests
Guardian	Yes	Liberal	Open	Global	Neutral

This evaluation of the strategies of relevant Media Groups provides us with the following understandings of the biases which underpinned some of the coverage of the events at Ferguson. By examining the "Biases and Editorial Bias," of the media group, its inherent "Point of View," the article outlines the "Type of Coverage" each group provided during the weeks between August 2014 and January 2015.

NBC News coverage could be categorized as Conservative, with a "Closed" point of view that wasn't flexible or understanding of the Ferguson community. This provided a "Top down" perspective with a "Neutral" or mainstream form of coverage which wasn't Investigative. CNN's coverage was more Liberal, with an "Open" point of view while still maintaining the "Neutral" or mainstream approach of most national broadcasters in the United States. From a print media perspective, *The Washington Post*, also provided a "Mainstream" view as a national newspaper, but maintained an "Open and Flexible" coverage of Ferguson's community.

The New York Times (2014) also maintained a "Mainstream" approach, but with more of an "Open" Editorial stance. Online coverage was more flexible, with the *Huffington Post* providing "Liberal and Open" coverage which was, at times, Pro-community and pro-protesters.

AlterNet and *the Daily Kos* provided much more "Radical, Organic" and "Bottom up" or community-orientated coverage, which were primarily Pro-protests. Fox News was, predictably, Conservative with very "Closed" perspectives, including authoritarian or "Top down" views which were anti-community and Anti-protests. Finally, *the Guardian* provided insightful analysis, with a "Liberal and Open" perspective which reflected their Global vantage point to an extent.

Through this survey, this article has presented an analysis of both mainstream and online coverage of the shooting of Michael Brown in Ferguson in August 2014. By exploring the reports and coverage of the shooting, protests, and Grand Jury, the report has highlighted the ideological biases of the media groups explored. It is not a comprehensive account of all media coverage, nor does the article cover the national civil rights campaign which followed the shooting in Ferguson and the Grand Jury decision not to prosecute the chokehold killing of Eric Garner by members of New York Police Department, and the subsequent killing of two New York Police officers in December 2014. Other incidents, like the shooting death of 12 year old African-American child Tamir Rice by a white officer in Cleveland in November 2014, have been the subject of in-depth investigations by the Department of Justice and also need further analysis. The rise of the "Black Lives Matter" social movements is a grassroots response from a community which has seen multiple deaths of unarmed African-Americans, often at the hands of over-zealous white police officers who seem to be lacking in conflict resolution training. The list of those who have died in such circumstances since the death of Michael Brown is disconcerting; 2015: Freddie Gray (Baltimore, MD) 2015, Walter Scott (North Charleston, SC) 2015, Tony

Robinson (Madison, WI) 2015, Anthony Hill (Chamblee, GA) 2014, Akai Gurley (New York, NY) 2014, Tamir Rice (Cleveland, OH) 2014, and many more (agabond.wordpress.com).

These events are the subject of a national debate on policing and civil rights. However, without doubt, a wider debate on these events is needed at national level. As the media is often the filter for our understanding of these events, it is important to analyze media coverage in conjunction with such pressing issues.

NOTES

1. "The Ferguson riots may seem similar to those in UK in 2011 – but there are stark contrasts" Newburn (2014).
2. Although Martin was shot by George Zimmerman, a security officer, the shooting added to a growing perception of oppressive policing within the African-American community.
3. According to the information gathering website 'Political Fact' Fox News gave false news 60% of the time (http://www.politifact.com/punditfact/article/2014).

REFERENCES

AlterNet. *AlterNet Online*. Ferguson coverage. Retrieved from http://alternet.org. Accessed on August 2014–January 2015.

Castells, M. (1998). *End of millennium, The information age: Economy, society and culture* (Vol. III). Cambridge, MA: Blackwell.

CNN News. Ferguson coverage. Retrieved from http://cnn.com. Accessed on August 2014–January 2015.

Daily Kos. Ferguson coverage. Retrieved from http://dailykos.com. Accessed on August 2014–January 2015.

Fox News. Ferguson coverage. Retrieved from http://fox.com. Accessed on August 2014–January 2015.

Gramsci, A. (1982). *Selections from the prison books*. London: Lawrence and Wishart.

Herman, E., & Chomsky, N. (2002). *Manufacturing consent: The political economy of the mass media*. New York, NY: Vintage/Random House.

Huffington Post. Ferguson coverage. Accessed on August 2014–January 2015.

McLuhan, M. (1994). *Understanding media: The extensions of man*. Cambridge, MA: The MIT Press.

NBC News. Ferguson coverage. Retrieved from http://nbcnews.com. Accessed on August 2014–January 2015.

Newburn, T. (2014). Comment is free. *The Guardian Online*, August 20.

Pew Research Center Survey August 14–17, 2014.

Powell, W. (2014). The roots of violence in Ferguson. *The Atlantic*, August 2014.

Szasz, A. (1998). *Ecopopulism*. Minneapolis, MN: The MU Press.

The O'Reilly Factor" Season 2014/8/24, *Season 8 Fox News Network*.

The Guardian. Ferguson coverage. Retrieved from http://guardian.co.uk. Accessed on August 2014–January 2015.

Washington Post. Ferguson coverage. Accessed on August 2014–January 2015.

SPATIAL PATTERNS OF MASS SHOOTINGS IN THE UNITED STATES, 2013–2014

Dinur Blum and Christian Gonzalez Jaworski

ABSTRACT

There were 374 mass shootings in the United States between 2013 and 2014 and they were not random occurrences. We expanded the FBI's definition of mass shootings (four or more injured or killed, not including the shooter), and collected data on every mass shooting that occurred in the United States to observe geographical patterns. Social Disorganization theories state that violence will occur in areas with high levels of poverty, large population density, and little economic opportunity. These theories work well with strain theory, which suggests that blocked goals, the introduction of a negative stimulus, or the removal of a positive stimulus leads to negative emotions, and these emotions can lead to crime in order to resolve the strain felt. From under this framework, we discover with point pattern analysis that there are patterns in the location of mass shootings crime scenes. They are not random. These crimes were far more likely to occur in the South, the Upper Midwest, and in Southern California, while they were considerably less likely to occur in the Pacific Northwest, Upper Midwest, or the Northeast. The lack of random spatial pattern indicates that the structural factors that

Environmental Criminology: Spatial Analysis and Regional Issues
Advances in Sustainability and Environmental Justice, Volume 20, 57–68

ISSN: 2051-5030/doi:10.1108/S2051-503020170000020004

contribute to mass shootings are more prevalent in certain areas of the United States than in others.

Keywords: Spatial patterns; mass shootings; United States; media; guns; public

This chapter focuses on a spatial analysis of mass shootings that occurred in the United States in 2013 and 2014. Based on an expanded definition of mass shootings, we found data on 374 mass shootings that fall under the expanded definitions that occurred in 2013 and 2014 in the United States. With the geographical information that we have obtained, we are able to graph each mass shooting incident on a map of the United States. Using point pattern analysis, we show the locations of mass shootings are not random.

We base our definition of mass shootings on the FBI's definition of the act, but alter it slightly. The FBI defines a mass shooting as having four or more people killed, not including the shooter. These events take place at a single location, and there is no distinctive pause between the shootings (i.e., these events happen at one place at one time, rather than being staggered over a period of time). We amend our definition of mass shootings to being a combination of four or more people killed or injured, not including the shooter. This allows us to include cases where a shooting took place, with some victims surviving.

Due to the comprehensive data we compiled, we are able to determine that there are three basic types or patterns of mass shootings. The most common type of mass shooting occurs at a house party or bar, and involves alcohol. In these cases, typically the victims and the shooters have been involved in some form of argument or confrontation. The second most common shooting is when the victims are either the significant other of the shooter or are otherwise close relatives of the shooter. The least common type of mass shooting is those that occur in public places where the victims have no connection to the shooter. However, the mass shootings that garner the most media attention are precisely those in which the victims and shooter did not know each other, while the more common mass shootings are either ignored nationally or are framed as gang violence, bar fights that got out of hand, or as domestic disputes.

Spatial analysis allows us to examine geographic clusters of mass shootings. This analysis allows us to see whether there are social trends in what is considered to be an individual event. Geographical patterns of these events force an explanation not centered entirely on the individual. While there are undoubtedly individual level factors that lead to every mass shooting, the geographical patterns that we discovered indicate larger social forces that are promoting mass violence. We examine where these shootings occur and find distinctive patterns in motivations for these events based on locales. Since there are patterns in regional variation of mass shootings, we attempt to explain why these events occur in some areas at much higher rates.

THEORETICAL BACKGROUND

For this analysis, we use both social disorganization theories and anomie or strain theories. These theories operate at different levels of analysis, with social disorganization being a macro approach focusing on communities, and anomie or strain being micro and focusing on individuals. However, one can embed anomie within social disorganization, as being embedded in poverty, or having crime be a regular, visible feature of life can act as a source of strain.

SOCIAL DISORGANIZATION

Social disorganization theories are rooted in Quetelet's idea of social cartography of crime, in which geography is emphasized rather than biology. Entering the 20th century, individualistic theories of crime were widely accepted, but these theories were myopic, as they focused on the "dangerous classes," meaning that people who were already labeled as criminals were the ones being studied, rather than it being a random sample of people (both criminals and non-criminals).

The focus of social disorganization is on traits of neighborhoods rather than individuals, as people live and interact with one another in social settings. This is a macro-level approach toward criminology, as one studies ecological units – neighborhoods, cities, counties, states, or nations, and focuses on the structural antecedents of crime: poverty, residential instability, and the prevalence of broken homes. These factors combine to disrupt core social institutions such as the family and schools – It is tough to establish a sense of community and continuity if the population is changing constantly and if there are few resources to invest in the community.

Early social disorganization scholars include Park, Burgess, and McKenzie (1925), who wrote that urban areas grow through a process of continual expansion from inner core (center) toward outer areas. As this process matures, cities have a central business or industrial area, and close to it is a "zone in transition," where poor newcomers settle to be close to factory jobs and inexpensive housing. Three more zones exist outside the inner city – workingmen's homes, residential zone, and commuters' zone (as one becomes established, one leaves the zone in transition to one of the other three zones).

Shaw and McKay (1942) built on this idea and focus on the intersection of persistent poverty, rapid population growth, heterogeneity, and transiency combine to disrupt core social institutions of society such as the family to have social disorganization. They found that over time, rates of crime by area remain stable, regardless of which ethnic group resided there, suggesting that there is something about the area rather than its inhabitants that affects delinquency. They focus more on affluent urban communities and poor urban communities, and contend that in more affluent communities, "the similarity of attitudes and values as to social control is expressed in institutions and voluntary associations designed to perpetuate and protect these values" (1942, p. 171). They argue that in areas wracked by poverty and constant social change, conventional institutions such as schools and churches become weak and a system supportive of crime is nurtured. We argue that this cycle is also related to money, as public schools in the United States are funded by property taxes. In areas where more people own homes, people pay more in property taxes, leading to stronger schools. In areas where more people rent, rather than own, there is less money to be invested in schools because fewer property taxes are paid.

Further, even in disorganized inner-city communities, parents and other adults try to teach children conventional moral values, but they have to compete against criminal influences that are not present in the more affluent communities, and these influences are tough to uproot. Once delinquent traditions take hold/become entrenched in a community, they are transmitted from one generation to the next, typically through peer groups. Sampson and Groves (1989) expanded on Shaw and McKay that areas with weak social controls would have high crime rates; they took for granted that the gap in between structural conditions (such as poverty and population density) and illegal conduct was social disorganization.

Early social disorganization studies emphasized structural conditions, but portrayed them as being constant and unchanging. These studies raised the question: What happens on to make people break the law at a high rate? Sampson and Wilson (1995) contend that race-based inequality in urban areas fosters the breakdown of the conventional institutions and cultural values needed to restrain criminal conduct. The cost of this inequality is borne most fully by African Americans, who must live in communities where 1 in 21 black males will be murdered, compared to whites where the rate is 1 murder in 131. Sampson and Wilson argue that culture is the acquisition of cognitive landscapes – If you are in an area in which violence is used to resolve disputes, then you might accept it as a viable way of resolving issues. As such, they attempt to build on the structural conditions by incorporating the variable of culture, which is rooted in these conditions. They argue that social isolation – "the lack of contact or of sustained interaction with individuals and institutions that represent mainstream society" – is what explains the presence of a culture that allows for higher crime rates. They write that in disadvantaged urban communities, youths live in segregated housing, attend schools in which virtually every student is a minority, and rarely travel outside the boundaries of their immediate neighborhood – They are cut off from routine activities that kids in more affluent areas witness, take for granted, and participate in.

Moreover, social isolation is not a bad choice made by inner-city residents. Isolation is the result of persisting racial inequality, which is the product of conscious political decisions (allowing racial discrimination in home purchases and ghettoizing minorities in high-rise public housing in geographically isolated areas) – and of broad macro-sociological changes (massive movement of jobs out of the inner city). Structural forces have isolated minorities in neighborhoods marked by extreme poverty and social disorganization, and they have cut off residents from mainstream society.

Sampson, Raudenbush, and Earls (1997) elaborate social disorganization by examining rates of violence across 343 Chicago neighborhoods. They show that "concentrated disadvantage" – a combined measure of a community's poverty, race and age composition, and family disruption – is related to neighborhood rates of violence; the effects of concentrated disadvantage are largely mediated by the degree of collective efficacy (willingness of community residents both to exercise informal control and to trust and help one another) in the neighborhood. They add to Shaw and McKay's idea of presence of control the notion that neighbors mutually trust/support one another. Social disorganization theories complement both conflict and control theories. After all, a major premise in conflict theories is that crime and delinquency are rooted in material conditions, and control theories assume that the reason people don't commit crimes is because there are controls – both externally imposed (either neighbors helping one another informally or the police formally) and internally socialized.

The strength of social disorganization theories is that they attempt to situate the individual in a larger social context. When culture is introduced as a variable, researchers are quick to show that it is rooted in material conditions, rather than leaving culture as an ambiguous variable that is borne by choice alone. Additionally, this group of theories can be complementary to several other groups of theories, such as conflict, control, and social learning theories. After all, the communities studied tend to be economically deprived, which allows for a culture that rewards (or doesn't punish) the use of violence and property crime. This culture is learned and taught. Further, in these communities, informal social control may be low, not necessarily due to acceptance of crime, but due to other factors. For example, if one is working multiple jobs to pay the bills, then they are not likely to be at home to check on their neighbors or on their own homes. Additionally, this group of theories has good policy implications. If community disorganization is the main source of delinquency, then the solution to crime is to better organize communities by introducing conventional institutions that support social control and funding them so that they are operational.

However, there are some weaknesses in social disorganization theory. Typically, the social context is assumed to be static and not dynamic, suggesting that once an area is poor it will stagnate and remain poor. Further, these theories focus on inner-cities, and do not allow for the possibility of social disorganization outside of these areas. In doing so, it assumes that wealthy areas will remain wealthy, and the same poor areas will remain poor.

ANOMIE AND STRAIN THEORIES

Strain theory is an attempt to explain crime separately from control and social learning theories, yet is complementary to both. In his foundations of General Strain Theory, Agnew focuses on three types of strain: actual or anticipated failure to achieve positively valued goals; actual or anticipated removal of positively valued stimuli; actual or anticipated presentation of negatively valued stimuli. This theory focuses on negative relationships with others: The individual is not treated by others as they want to be treated. Adolescents are pressured into delinquency by negative emotional states (e.g., anger, frustration) that often result from negative relationships. As such, negative emotions are key to understanding crime, much as social isolation and culture are to social disorganization theory.

Much of the work done in strain theories focuses on adolescents. Agnew writes, "adolescent strain is typically measured in terms of the disjunction between aspirations (or ideal goals) and expectations (or expected levels of goal achievement)" (1992, p. 51). This is one of the social psychological components to strain; another such component is the equity component, which states that people who are under-rewarded will feel strain more than those who get their fair share; this feeling of being under-rewarded can lead to delinquency or crime to increase rewards (i.e., theft), decrease inputs/effort (i.e., truancy), lower outcomes of others (theft again, vandalism); or increase inputs of others (being disorderly). In short, a person who feels they have been shortchanged for their efforts can resort to negative behavior in order to alleviate the strain they feel. This negative behavior can include violence and crime.

Agnew (1992) argues that "strain stems from the inability to achieve certain ideal goals emphasized by the (sub)cultural system" (1992, p. 51). He argues that there are coping mechanisms that people can employ to cope with strain that do not involve crime, but that those who turn to delinquency or crime may be unfamiliar with the legitimate coping techniques. Some individuals turn to delinquency due to not having coping resources, conventional social support, or that macro-level variables reward delinquency more than conformity.

For Agnew (2001), the strains most likely to result in crime are viewed as unjust, seen as high in magnitude, recent, of long duration (2001, p. 323), associated with low social control, create pressure to engage in criminal coping (2001, p. 326). Additionally, strains can be subjective (disliked by the specific people experiencing them) or objective (when the event is disliked by a large group of people). Further, it is the cumulative effect of strains and stressors that the individual feels that matters more than any specific stressor.

Strain theory allows for social psychology and emotion to enter our consideration of why people might turn to crime or delinquency, and given that unjustness is a source of strain, it suggests that mechanisms can be put in place (in schools or workplaces, for example) to allow for people to have a say in how resources should be distributed and for grievances to be aired. Further, it emphasizes that interpersonal contact is important, and that a lack of respect, warmth, kindness, or love can be a source of strain.

There are some limitations to strain theory. There is a focus (almost exclusively) on monetary success or attaining middle-class status. While this can explain crime as a response to strain born from poverty, it does not explain middle-class delinquency. Further, it ignores barriers to goal achievement other than social class and offers a fairly weak explanation as to why only some strained individuals turn to delinquency or crime. Additionally, these theories focus on adolescents and do not examine adult criminals.

Broidy and Agnew (1997) offer explanations for the gender gap in crime. The gap is highest for serious violent and property crimes, while being lowest for property crimes, drug use, and escapist behavior like running away. They argue that males and females experience different types of strain, with male strains being more conducive to serious violent and property crimes. These strains can be financial, which is conducive to both property crime and severe interpersonal conflict, which is conducive to violence. Effectively, poverty can lead a person to try and get money through robbery, selling drugs, prostitution, and possibly violence. Females are also subject to family violence, which might explain why gender differences in commission of family violence are relatively small. This strain would be found in abusive relationships in which a woman kills her abuser.

Broidy and Agnew also write that males and females differ in their emotional responses to strain. They find that female anger is accompanied by depression, guilt, anxiety, and shame, while male anger is accompanied by an amplification of the anger, and can result in violence. They argue that there are differences in coping, social support, opportunities, social control, and disposition to engage in crime can lead to males being more likely to respond to strain with serious property or violent crime.

DATA AND METHODS

We created a database of every mass shooting that has occurred in the United States from 2013 to 2014, using a combination of police reports and news reports. Our data include the addresses, zip codes, and geographical coordinates of every mass shooting during this two-year time span. The exact geographical location of all 374 mass shootings was graphed on a map of the United States.

We performed a point pattern analysis to understand if there are patterns among the locations or if these events occur at completely random places, and whether any geographic patterns found are statistically random or not. We found several geographic patterns to mass shootings. These mass shootings occurred far more often in the South, the upper Midwest, and California. There were very few in the Mountain West or the Upper Pacific Coast.

The null hypothesis for our analysis assumes complete spatial randomness. Within the boundaries of the United States, the authors test if there are any patterns among the locations of the 374 mass shootings. If the data is clustered, the null hypothesis will be rejected and the assumption that the locations of the mass shootings are not random.

A point pattern analysis was performed for all of the instances of mass shootings in the United States from 2013 to 2014. First, the mean geographical center was plotted. Afterwards, the distance from the location of each mass shooting to the geographical center is measured. Since there were no mass shootings that occurred in Alaska or Hawaii, we were able to exclude those long distances from our analysis thus concentrating only on the lower 48 States. From the distances of each shooting, a standard distance deviation was calculated measuring each point from the mean geographical center.

The conclusions of the point pattern analysis indicate that the data is not randomly dispersed and is indeed clustered in patterns. Sufficient evidence was found to indicate that there is indeed clustering of the incidents of mass shootings. The null hypothesis is rejected thus the assumption of randomness is rejected. The data is visually clustered in the South, the upper Midwest, and in California (Fig. 1).

2014 ALL AGES IN POVERTY

Fig. 2 shows poverty rates in the United States in 2014, with darker colors indicating a higher poverty rate (32.6–52.2% of the population living at or below the federal poverty line). We found that by state, poverty mirrors the states with the most mass shootings, suggesting that poverty may play a role in these events.

FINDINGS AND DISCUSSION

During this two-year span, there were 374 events that fell into our expanded definition of mass shootings – a rate of almost one mass shooting every other day. We found concentrations of mass shootings in Southern California, the southern and southeastern United States, and some clustering in the northeastern United States as well as the Upper Midwest. Mass shootings are clustered in areas that are poor, and where there are large concentrations of people. What we also find is that alcohol plays a role in these mass shootings. Most of the shooting events were bar fights or house party fights, when alcohol was

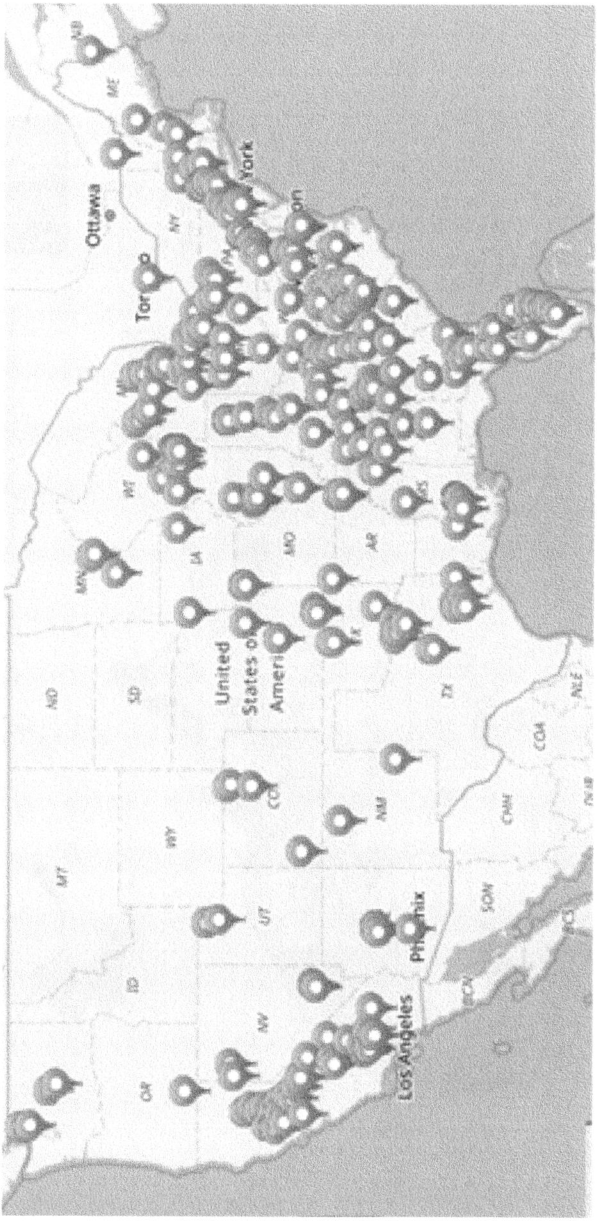

Fig. 1. Mass Shootings in the United States, 2013–2014.

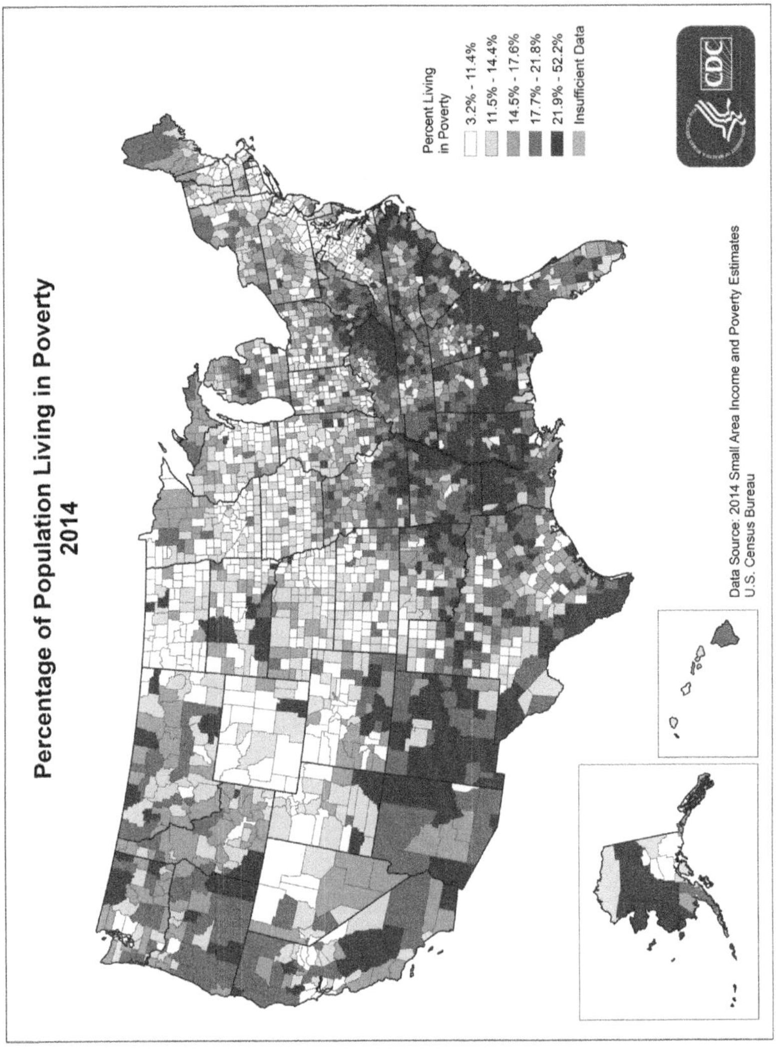

Fig. 2. Poverty Rates in the United States, 2014.

freely available. The combination of alcohol, an argument, weapons, poverty, and a lot of people in an area combined to make a lethal cocktail.

We argue that our spatial analysis lends support for both Social Disorganization and for General Strain Theory. Poverty acts as a constant strain, and violence may be a common occurrence as a way to cope with the negative emotions caused by the strain. Once the constant strain of poverty is supplemented with the acute strain of an argument that happens at a bar or party, and have alcohol to exacerbate tensions, the end result is a social environment ripe for violence. The arguments in and of themselves may have not been enough to spark a violent reaction, but in combination with alcohol as an immediate amplifier, and poverty as a constant source of tension, and these served as motives for most mass shootings.

We note that the higher-profile mass shootings (Aurora, CO; Newtown, CT; and later Santa Barbara, CA) differed from these, in that the shooters in the high-profile cases came from middle-class or upper-middle-class families (if not upper class in the Santa Barbara case), so poverty was not a motive for these shooters. However, we argue that in all cases, shooters felt that they experienced some sort of strain, be it chronic, acute, or both, and resorted to violence in order to resolve their strain. These suggest that the social problems of social density (high population areas) and competition for scarce resources (poverty) serve as contextual factors for lethal violence.

We suggest several possibilities for action to reduce these mass shootings events in the future. One idea is to limit alcohol sales when a person is visibly and/or audibly agitated. Given that the role of alcohol and arguments preceding a mass shooting event were positively and strongly correlated with each other, if alcohol sales are limited, it may reduce the anger in arguments, or may allow for cooler heads to prevail. We may not be able to prevent shootings at house parties, as alcohol is offered free of charge at these gatherings, but this would reduce shootings that occur at or outside bars.

The next suggestion is to reduce firearm availability. This is tricky, given the legal right to own firearms in the United States. However, we recommend a background check that not only examines mental health history, but also looks at a customer's criminal history (if applicable), and asks questions regarding previous histories of arguing or physically fighting. While reducing firearm availability does not address the cause(s) of these mass shootings, restricting their availability may reduce both the number of mass shooting events as well as the years of life lost to firearms, and years of life lost more generally.

Further research can be done to see whether poverty and mass shootings are linked by zip code, not just by state. If the two are linked, we would argue that poverty acts as a potentially lethal strain. Research can also be done to see what patterns based on race, class, and gender of offenders exist, and whether the same patterns of shootings continue or change. We suggest that mass shootings be understood as reactions to social pressures, both macro and micro, and are not solely psychological disorders (Figs. 1 and 2).

REFERENCES

Agnew, R. (1992). Foundation for a general strain theory of crime and delinquency. *Criminology, 30*(1), 47–88.

Agnew, R. (2001). Building on the foundation for a general strain theory: Specifying the types of strain most likely to lead to crime and delinquency. *Journal of Crime and Delinquency, 38*(4), 319–360.

Broidy, L., & Agnew, R. (1997). Gender and crime: A general strain theory perspective. *Journal of Research in Crime and Delinquency, 34*(3), 275–306.

Park, R. E., Burgess, E. W., & McKenzie, R. D. (1925). *The city: Suggestions for investigation of human behavior in the urban environment*. Chicago, IL: University of Chicago Press, 1967.

Sampson, R. J., & Groves, W. B. (1989). Community structure and crime: Testing social-disorganization theory. *American Journal of Sociology, 94*(4), 774–802.

Sampson, R. J., Raudenbush, S. W., & Earls, F. (1997). Neighborhoods and violent crime: A multi-level study of collective efficacy. *Science, 277*(5328), 918–924.

Sampson, R. J., & Wilson, W. J. (1995). Toward a theory of race, crime, and urban inequality. In J. Hagan & R. D. Peterson (Eds.), *Crime and inequality* (pp. 37–56). Stanford, CA: Stanford University Press.

Shaw, C. R., & McKay, H. D. (1942). *Juvenile delinquency in urban areas*. Chicago, IL: University of Chicago Press.

HOMELESS DEMOGRAPHY IN LOS ANGELES COUNTY

Hugo Aguas

ABSTRACT

This is a demographical exploration of a wide variety of topics, which are as follows: gender, race, age, employment, substance abuse, mental illness, physical illness, veteran status, government assistance, physical & sexual abuse, hunger, and space. All of these topics were explored in conjuncture to ascertain who the homeless are. To explore this topic, data from LAHSA (Los Angeles Homeless Services Authority) was utilized to calculate demographical aspects of the homeless population with a raw sample of 4,852. I coded this data to further find insight among the population. Throughout this study it was found that nearly 60% of the homeless population in Los Angeles County are unemployed, 50% of the population have been incarcerated, a third of the population is homeless by age 20, a quarter of the population are women, and a third don't utilize government assistance programs.

Keywords: Homeless; demography; Los Angeles; marginalization; deviance; poverty

INTRODUCTION

This chapter regards demographics of the homeless population in Los Angeles County. In addition, it is fair to note that this work is inspired by personal

Environmental Criminology: Spatial Analysis and Regional Issues
Advances in Sustainability and Environmental Justice, Volume 20, 69–86

ISSN: 2051-5030/doi:10.1108/S2051-503020170000020005

experience. I recall the first time I witnessed the tents on my way to the center of Los Angles; the occupants were methodically fixated on camping out on the side of road, while others aimlessly wandered or resigned themselves to sleeping on the floor. The unfortunate souls who were unable to obtain a tent would sleep on cardboard boxes or on the solid concrete with only the clothes on their backs to protect them from the elements. Entering Skid Row near the Fashion District was an out of body experience. Accepting the extensive reality of homelessness had a profound effect on me and my quest, since that day my main prerogative is to discover who they are.

This study is meant to elaborate on the homeless paradigm. This is to be done through sheer basic demographical research and an elaboration on employment and government assistance use. The aforementioned topics are concepts of interest that will be incorporated into my thesis work. As for this chapter, the article will entail topics such as: gender, sexuality, veteran status, space the homeless occupy, relationships, sexual/physical abuse, rate of homelessness, situation prior to homelessness, employment and jail/prison associations, forms of income, employment status, employment and physical/mental disability status, government assistance, ER utilization, employment and veteran status, employment and mental illness, substance/alcohol dependency and employment, criminal backgrounds, alcohol abuse, drug abuse, and age of the first time homeless. All these topics are defining characteristics of a typical subset of the homeless population. The vast array of topics mentioned were chosen to unveil the population and while these are not defining characteristics, they must be understood at face value before one can make sense of the population or dwell on their behavior.

LITERATURE REVIEW

The survey I originally distributed contained a large variation of questions that have been posed by the LAHSA survey. The purpose of this literature review is to understand the homeless population to conceptualize the framework of the data to be analyzed. In addition, this literature review will compile a detailed synopsis of five different articles ranging from a wide variation aspects concerning gender, social services, veteran status, homelessness leisure activities, assessed risk factors in relation to homelessness, and public policy on skid row homelessness; throughout my study, the aforementioned and other articles will be discussed. The first article under review is based on the sexualization of homeless women.

In the first journal Jenifer Wesely, the researcher, conducts qualitative interviews to talk about the gendered experience of homelessness women. She notes the reinforcement of femininity through culture, which inadvertently turns into sexualization. Wesely takes a simple interactionist approach to define how women wind up being sexualized throughout the homeless women's lives. The gendered power structure is the epitome behind the abuse of female behavior. Behind the façade of gendered expression, homeless women face further disadvantages such as "the institutional effects of structural gender inequalities on women, including inadequate education, poor earning power, limited job opportunities, overwhelming child care responsibilities, and fragmented support networks as well as patterns of physical and sexual abuse" (Wesely, 2009, p. 92). Wesely highlights that many homeless women face high rates of sexual abuse as children compared to their male counterparts.

The method of this derived research came from one-on-one qualitative interviews based upon "the structural conditions of homelessness and gender inequality on an individual level, through the lived experiences of the women participants" (Wesely, 2009, p. 93). The topics the researcher discussed about, with the interviewee, consisted of economics, stigma, and lack of educational/work opportunities. The women interviewed were first asked about their childhood in relation to child abuse. Many noted that they were sexually degraded early on which led to a new construction of the self. Wesely further noted that her participants' surroundings "were further compounded by growing up in multiple disadvantaged families, having few role models or social outlets early on, and lack intervening supports and resources, framed meanings on homeless life and survival contexts for the women interview" (Wesely, 2009, p. 93). The researcher notes these interactions to promote a linear storyline of her participants and their daily social structure.

The sample size of this study is quite small, being only 20 women who resided in a homeless shelter in an undisclosed Southeastern city. Case managers of certain homeless facilities asked only 20 women and all the individuals decided to participate in the survey. Each participant was given $10 after the interview, which was typically lasted 2 hours. The survey was taped, with permission, to compile key concepts for progressive analysis. The demographics of the study consisted of six African-Americans, three Hispanics, one mixed African-American/Native-American, one mixed Puerto Rican/African American, and nine Caucasian women. The age range of the participants consisted a minimum of 19–62 years of age. She notes that she identified themes from the interviews and was then able to code them. The topics, she noted, consisted of "background and childhood, family relationships both as a child and as an adult, intimate relationships, power and control, violence abuse, and homelessness" (Wesely, 2009, p. 94).

Wesely breaks down the scope of her study into three different sections. The first is degradation of the female, the second is fear and vulnerability, and finally survival as a woman while being homeless. And for the degradation of the female, Wesely noted that their lives began during adolescents. Many immediate family members reinforced the notion of looking down upon their female child, for example some respondents noted "hearing their mothers and father about how women were for only one thing, and that one thing did not count for much" (Wesely, 2009, p. 95). The respondent was degraded verbally and their personal worth was validated as a sexual object. Some mothers even supplied their children with pointers on how to attract males, after all that was their perceived worth. Overall, out of all the women "sixteen of the twenty women endured some sort of physical or sexual victimization as children, and all experienced emotional abuse or neglect" (Wesely, 2009, p. 96). These women were ergo socialized into their social paradigms.

The next topic is vulnerability and fear. Wesely argues that women feel physically vulnerable to society due to their beliefs of self-worth. Therefore women, who are often captivated "to adopt a tenacious commitment to caring for anyone who promises love, material success and acceptances, such that it represents an extreme liability for self-survival" (Wesely, 2009, p. 97). This promotes a problem, because the certain male caretakers tend to be abusive individuals who provide minimal comfort through immediate acceptance. Wesely noted a pattern in one of her subjects; her subject noted that all her relationships have been physically abusive, but the experiences are associated to her childhood victimization. Another experience was the institutionalized force that places women in further situations of harm. Another individual once called the police to resolve a domestic abuse situation, but due to the excessive calling of the respondent in the past, the police told her if they arrived, someone would be going to jail. The problem was that the husband always left home before the police arrived to resolve the situation. Aforementioned situations depict the institutionalization that undermines a woman in a patriarchal society.

The third frame Wesely highlights is homelessness and survival. Due to women being objectified and painted as easy targets since childhood, the problem is then transferred onto the street. Wesely notes that "Women report constantly monitoring their environment for signs of danger, hesitating to venture outside alone or even in the company of other women, asking men for protection, modifying their clothes … and restricting their activities" in order to even get by in the streets (Wesely, 2009, p. 100). Their objectified bodies at certain points even turn into a survival tool. One individual states sometimes having sex as a way of obtaining food, shower access, and clothes. In a separate form of female solidarity, some women band together at night in order to protect themselves against male predators. This gendered review on the female homeless does have certain limitations. The sample is tiny, which means the study may have weak external validity. Another problem is that all the women are self-identified abuse victims, which means their responses cannot be fully documented. Finally, the researcher notes that these were extreme cases and may not fit the paradigm of all homeless women.

The second article concerns social service perception by the homeless population. The research was conducted via qualitative means by asking the respondents about their experience within the social service system in Portland, Oregon. The population consisted of over 515 participants and was conducted by the Sisters of the Road, which is a non-profit organization. The information was then coded into a variation of themes ranging from "employment to services to self-esteem" (Hoffman & Coffey, p. 209). The themes of the results obtained were calculated as positive conditions of social services, negative conditions of services, positive staff experiences, and negative staff experiences. Many homeless individuals have described their experience with social service as not being "treated as fully recognized adults or respected as equal citizens, but rather as numbers and children" (Hoffman & Coffey, 2008, p. 208). This is a problem, because it disregards the view of the self and many homeless individuals are systematically forced to flock to these social service institutions without full psychological comprehension or compassion.

The study depicts social service workers as being "constrained by limited resources, forms of bureaucratic authority, professional identities, low wages, fear of losing their status and privileges, and perhaps even precarious and abusive working conditions" (Hoffman & Coffey, 2008, p. 211). Social service workers may be unintentional in their behavior toward the homeless, but the external forces pressure them into their mode of bureaucratic authority. This authority stricken behavior has led to certain homeless people avoiding social services to resist being undermined within an already constrained system. The following concept concerns the infantilization and objectification of the homeless by social services.

A significant proportion of respondents noted that they had experienced negative conditions of social services and negative staff experiences. This is largely derived from the objectification of the homeless. Certain homeless individuals believe social services doesn't care about the conditions of man due to their living situation. For example, "negative comments about interactions with staff, for instance, were overwhelmingly about issues of disrespect, rude treatment, and lack of compassion, incompetence, and unethical behavior by staff" (Hoffman & Coffey, 2008, p. 213). Many respondents complain that are treated like children because they have lost their ways, which leads respondents to the feeling of being disrespected and dehumanized. Those who experienced an extreme form of disrespect are those "new arrivals and those without employment," which was coded with a red dot (Hoffman & Coffey, 2008, p. 213). Due to the lack of preferential treatment, many opt out of social services, which then creates more issues such as "homeless camps" (Hoffman & Coffey, 2008, p. 213). The authors then promote the concept of saving or salvaging the self from institutional stigma.

The homeless population is a highly stigmatized population and many don't want to be consumers of social services. Therefore, in order to preserve the notion of the self, certain portions of the population will opt out. These individuals believe in this "sense of independence and often express disdain for people who did not opt out" (Hoffman & Coffey, 2008, p. 216). Not accepting social services is a form of rebellion which preserves the conceptualization of the self. It relies on inner-dependence to survive in lieu of being objectified in order to receive basic services.

Not all experiences were negative, although the positive responses were a small percentage compared to the negative ones. Another fascinating aspect is the heavy internalization of negative experiences. Those who comment on positive experience don't usually comment as vividly compared to the negative ones. Individuals who were fully respected and dignified mentioned that they were accommodated appropriately and they positively highlighted the experience. There is a gendered notion in experiencing respect. The researcher noted that the gendered experience benefits women because "women, as opposed to homeless men, are potentially rehabiliatable" (Hoffman & Coffey, 2008, p. 217). This translates into the belief that women are fully capable of recovery, while men are treated discriminately for their gender. Positive experiences also derive from services that allow the homeless individual to work for a specific goal such as a meal. Work manages to uplift the notion of the self. The following study under observation concerns risks of homelessness in relation to veteran status.

This study was conducted by the CDC and the title regards the full scope of study, being "Prevalence and Risk of Homelessness among US Veterans." The data from the survey concerns 130,554 homeless individuals with the data deriving from seven jurisdictions and homeless service centers. The study notes that "Veterans are overrepresented among the homeless in the United States and are at greater risk than nonveterans of becoming homeless" (Fargo, 2012, p. 1). The researcher notes that in 2009, there were 136,000 homeless veterans with 53% having a severe health condition (Fargo, 2012, p. 1). The question the researcher aims to answer is as follows: is there a correlation between veteran status and an increased risk of homelessness? The second question is, does veteran homelessness vary in regards to age, race, or sex?

The first part of the study discovered that out of 130,554 individuals receiving homelessness services, 10,726 were veterans. The researcher found that "compared with nonveterans, veterans in each population were disproportionally male and in the older age categories" (Fargo, 2012, p. 3). In fact it was found that in terms of gendered experience, the risk ratio of homelessness compared to non-veterans, were two to one, if you were a male and, three to one, if you were a female. A discovered trend concerning high risk populations associated homeless veterans who are likely to be young African-Americans.

The second part of the study found that "veteran status, older age and black race were significantly and indecently associated with risk of homelessness among both men and women" (Fargo, 2012, p. 3). In terms of gendered experiences, a high risk association was found between men within the age range of 45–54 for veteran homelessness, but oddly enough the number tends to taper off after 54. As for women, "risk for homelessness among women declined with age at an increasing rate in both the general and poverty populations, so that older women were at the lowest risk for homelessness, compared with the youngest group" (Fargo, 2012, p. 3). A peculiar association would be veteran status in relation to post-9/11 military service to describe the aging phenomenon.

Finally, African-Americans are disproportionately represented regardless of the study and general population.

The highest variable in lieu of homelessness would be access to health care and veteran status. A problem unrooted was the fact of "Given their (veterans) lack of housing and heightened susceptibility to chronic health problems, homeless veterans will likely contribute disproportionately to the increased demand for long-term care through the VA" (Fargo, 2012, p. 4). This depicts the high causal rates and standards that are associated with being a veteran, which makes the population susceptible to homelessness. The following piece of literature concerns the leisurely activities of the homeless in Las Vegas.

The following article details the homeless and their leisure activities in Las Vegas; in addition it is yet another qualitative article concerning certain individuals and their daily routines. The qualitative study was an ethnographic project based upon four dozen individuals. The question the researcher aims to answer is "If homeless people are viewed as more than just poor but as actively engaged in a lifestyle, I ask what activities beyond meeting basic survival needs (such as leisure activities) characterize that lifestyle?" (Borchard, 2005, p. 2). Leisure is described as free time in which activities are pursued. The conducted interviews were formulated between 30 minutes to approximately 2 hours. The main interviews Bouchard utilized were with individuals with whom he had multiple contacts. In the following sections, Bouchard describes the lifestyle of five individuals being Ron and Ricky, Dave, Jessi, and Jack.

Ron and Ricky were initially sitting at a Starbucks when Bouchard ran into them; he noticed the men using laptops at a Starbucks without any coffee products. When he approached the two African-American men, they struck up a conversation on homelessness and they admitted to being homeless themselves. The pair admitted to being recent homeless individuals who came to Las Vegas for career opportunities. The two individuals were wearing casual clothes, which allowed them to integrate with the general crowd. The two admitted they had been staying at a homeless shelter for the previous couple of weeks and they had recently combined their housing vouchers for a single bed hotel room. In addition to their housing trouble, Ron and Ricky noticed there were many barriers to entry for employment such as the need for a "TAM card, Sheriff's Card, Health Card" (Borchard, 2005, p. 7). Ron and Ricky noted other troubles concerning transportation. They noted finances were usually divided between paying for public transportation and having enough to eat. After the first day of conversing, Ron and Ricky would reappear within a couple of days. Throughout the associated time, Bouchard noticed a hobby that Ricky undertook was online gambling. Ironically enough, he would gamble while loitering at the Starbucks by utilizing the business's Wi-Fi. A separate activity the pair indulged in was people watching. They would spend time walking through the Strip to eye the tourists and their characteristics, such as social class to speculate how much money an individual owned. When not on the strip, these men watched DVDs, listened to music, or watched television at shelters. This exhibited an ethnographical trend within the homeless population, concerning the use of electronics. Shelters are a meeting grounds for other similarly situated individuals. At shelters, the homeless build efficacy with others. Dave may have a similar story, but the way he passes time differs from that of Ron and Ricky.

Dave outright noted that he uses technology and other forms of it to spend his leisurely time. Dave is a homeless panhandler from England. At age 55, he has accepted his living situation and has done so in a way to suit his everyday needs. Dave notes that he utilizes "the public library, the 99 cent store, and a community center (with showers and a recreation room)" (Borchard, 2005, p. 15). He uses the library to read and listen to the BBC on the computer, more specifically on a Real Player application. The 99 cent store is for basic necessities and at the community center he is able to take a shower. Unlike many homeless people, he enjoys being homeless and still believes in the freedom aspect of being without a residence. Technology has been such an integral part of the new wave of homelessness that Dave even utilizes a motion detector to protect his belongings from theft at night. The final individual is Jessi, and her story is the most astonishing of all.

Jessi is a Native-American who had been living in Las Vegas 6 months prior to meeting Bouchard. When they met, she was at a park in Downtown Las Vegas drinking with her cohort of friends. She is an educated woman, who was unable to find a job and fell into the world of homelessness. In addition, a prior traumatic incident (one that cost her leg) had rendered her ability to move forward in life as she stigmatized herself to the bottom. Adding to her misfortune, "she says she dishonored her parents when she left home with Terrence because he was black" (Borchard, 2005, p. 18). Due to all these traumatic experiences, she spends time lingering around with friends and drinking. She engages with these individuals because they share "social status of those in the park, their marginalization from other friends and family, and their current social conditions might have led them to seek out others in a similar position for acceptance and made those immediate friendships important" (Borchard, 2005, p. 21). Finally, there's Jake.

Jake is an elderly African-American man, who also happens to be a veteran. Jake spoke of being a common visitor to shelters, especially during the summer; sometimes just to relax and watch some movies. At night, he unfortunately has to worry about raids on the homeless community; often, he notes, there are raids on the streets to move the homeless population on grounds of looting, but recently the law has been ruled as illegal. He is a reformed drug addict and he acknowledges the use of his money is invested into video poker machines. When he is not day laboring at the machines, he notes his enjoyment of attending the library. There he can regroup and read. Ultimately, the researcher noted that the homeless "engage in leisure pursuits for the sake of enjoyment or simply distraction from their problems" (Borchard, 2005, p. 24). The next essay concerns public policy within Los Angeles's own Skid Row.

This analysis was conducted by Dr. Culhane and in it he attempts to discover the differing policy strategies in the city of Los Angeles. The city automatically deploys laws "to criminalize homelessness with ordinances that target panhandling, loitering, and sleeping in public" (Culhane, 2010, p. 852). The problem has been so widespread that certain vehicles are used just to transport the homeless to other sections of the city. The purpose of these proposed tactics were to enforce the Safer City Initiatives, which was aimed "to reduce crime in the downtown district of Los Angeles" (Culhane, 2010, p. 851). But the issue with the initiative is that instead of fixing the problem of loitering, the city temporarily displaced the homeless, who would end up going back due to the social services at Skid Row. The author then argues for more scattered social services, an end to harsh discrimination practices, a conclusion to ineffective and costly policing practices by LAPD, and more fundamentally, reinforced housing assistance programs for the homeless.

DISCUSSION

A discussion concerning the literature concerns the limitations they place upon my demographical study. A significant portion of the read literature relates to highly stigmatized populations within the homeless. These groups can consist of those with mental illness, physical disabilities, substance abusers, and veterans. While these are all topics that should be explored, this study has reached a point of information glut. On the other hand, this study will incorporate the aforementioned demographics and a variety of others. As will be noticed, this study devotes a considerable amount of time on employment, government assistance, and income. This research is intended to broaden the horizon on such literature and the emphasis that is placed is to support other researchers to do the same.

The articles that have been studied concern topics such as law, inhabited space, leisurely activities, sexualization, substance abuse, veteran status, mental/physically incapacitations, risk factors to becoming homeless. These topics were explored to assess the problems that are of most pertinent within the study of homelessness. Therefore, this study placed a considerable amount of emphasis on these topics, as will be noted in the findings. Articles, such as the aforementioned, are to be used in conjecture with my demographical study to highlight various concepts that plague Los Angeles County.

DEVIANCE

This study has been incorporated in a book concerning explorations in deviance due to the highly stigmatized nature of the population and their level of disaffiliation within a circumstantial portion of society. Therefore, this study of research regards the demographic components of the homeless population to describe their alienation. One thing that was not included in this study is theory. This study failed to incorporate theory due to the demographical nature of the work. In a future article concerning homelessness and employment/ government assistance use, the topic and theory will be explored and utilized.

METHODS

When it comes to my received data, it completely stems from the 2016 Greater Los Angeles Homeless Count Survey, which stems from LAHSA and is in conjuncture of assistance from the University of North Carolina at Chapel Hill. I came across this data set upon researching for statistics concerning the homeless population at a variety of homeless shelters (LAHSA, 2016). While I volunteered Downtown, I asked coordinators to direct me toward homeless statistics in Los Angeles and was recommended to LAHSA. After this point, I met with LAHSA's senior analyst and he prepared the raw data for me. Ultimately, the data consisted 4,852 face-to-face interviews, but was originally stratified from 2,053 census tracts to 409 randomly selected census tracts. The method in conducting the data is as follows:

"Analysis of the 2016 Greater Los Angeles Homeless Count (HC16). The CSRL's responsibilities were to provide methodology and process direction, to design and select all required samples, to develop estimation strategies, and to produce weighted survey estimates with relative standard errors where applicable. LAHSA's roles included managing all other data collection tasks, which included conducting the street and shelter counts, the contents of survey questions and forms, administering the demographic survey, compiling the estimates for presentation, and conducting a separate youth count" (Agans, Bowling, & Zeng, 2016, p. 4).

The derived data stems from LAHSA (Los Angeles Homeless Services Authority). The data comes from a large survey which counted the population by using field assistants. The population was offered a five dollar gift card upon completion and was completely anonymous to reinforce the internal validity. The external validity is highly generalizable based on the large population of the sample and the differing field workers. The study was in correspondence with city of Los Angeles, LAHSA, and the University of North Carolina-Chapel Hill. The work was meant to provide a representable sample of the entire homeless population in Los Angeles County and add to the longitudinal work collected previously.

FINDINGS

For the first section of this chapter, the topic under discussion will be race and age of the homeless in Los Angeles County. These statistics were worrisome and yet, unsurprising. It was found that African-Americans consisted of almost 41% of the entire homeless population. This population is grossly overrepresented in regards to their demographical makeup of the United States. Following is the Caucasian group, which compiled at 34% of the population. Next is the Hispanic population which totals to 26% of the population and American-Indian or Alaskan Natives compose of almost 4% of the homeless population. Asians composited for close to 2% of the entire population, while dead last came Hawaiians at under 1%. Age on the other hand did not vary. LAHSA reported that nearly 70% of the population was between 25 and 54 years of age, while, 20% of the population were 55–61 years old. These are individuals who have not yet reached retirement age, and yet they are living on the streets.

In regards to the second section of the compiled work, this section ascertains information on the first time an individual was homeless. This data notes that there is a steady increase in homelessness in age, but will typically taper off. Those who stated that they were homeless by 10 is at 1.8%, by 20 it's at 33.5%, by 30 it drastically increases to 64.8%, by 40 it's at 83.3%, by 50 it's at 94.6%, and is finally surmised at 77 with 100% of the population being homeless; with of course, 77 being the outlier. I hypothesize that with age, problems increase and adolescence is the most vulnerable an individual can be. Vulnerabilities that are to be considered should be drugs, negative childhood experiences, lack of positive reinforcements, and jail/prison. A fellow researcher notices a trend within gender. For example, the homeless women Wesely interviews notes: "All of the interviewees grew up in disadvantaged and marginalized conditions characterized by varying degrees of poverty, neglect, loss and dislocation, drugs, alcohol, and illness" (Wesely, 2009, p. 96). These levels of negative experiences may prove to be a prelude to homelessness. The following section concerns alcohol, drugs, and employment.

As noted, a homeless individual faces negative experiences from the get-go. The current population exhibited high rates of alcohol and drugs use. The ascertained data denotes that around 20.9% of the population abused alcohol, while another 20.6% abused illicit substances. To assure validity on this topic of interest, the researcher empirically evaluated the individual for drug or alcohol abuse and reported that 36.6% had traces of either substance/alcohol abuse. Such an abuse could lead to a loss in employment opportunities and an experimental study found that "within each treatment group, individuals who established early and length stretches of abstinence were most likely to be stably housed and employed at 12 and 18 months" (Milby et al., 2010, p. 917). It should be common knowledge that drug use leads to a low likelihood of employment due to instability, and this idea was reinforced by the data. The data found that less than 1% of general population had a full or part-time job and abused drugs. 13.1% had no job or looked and for alcohol the numbers are similarly startling, lying at steady rate of 12.7% for homeless individuals who abused substances and had no form of employment at the time.

Throughout this demographical overview, employment, income, and government assistance will be of high interest. The next part concerns general employment statistics.

Employment within the homeless population varies depending on the section. Those who worked full-time composed of 1.5% of the population, part-time constituted of 3.5%, seasonal follows at 1.1%, temporary work stops at 3.7%, and those who are unemployed and looking consisted of 35.4% of the entire population, the unemployed who were not looking follows at a high of 40.5%, and finally none of the above constitutes at 14.8%. The problem is that a large segment of the population is looking for work, but the homeless end up discouraged, which demoralizes them to avoid seeking out employment. The employment paradox is that "employment represents an important means of successfully exiting homelessness and should be a priority to end homelessness" (Poremski, Whitley, & Latimer, 2014, p. 181). Although many may want to exit homelessness, their current predicament incarcerates them into this vicious systematic cycle.

Employment isn't the only way to ascertain income. A variety of forms to attain capital include panhandling, recycling, day labor, and none of the above. The data found that the population resorts to panhandling 29.8% of the time, recycling follows at a strong 35.3%, and day labor under-performs at 10.5%. These are not surprising finds, although it must be noted that the resourcefulness of the population is unique. In another study, two men gain their income by gambling in Las Vegas. The author refers to this type of sustainability as "street entrepreneurs because prison records and/or periods of drug addiction made more traditional types of employment impossible" (Borchard, 2005, p. 9). In the survey, there was another category that listed off the ways individuals attained their income, unfortunately I did not have enough time to code all of the legal and illicit activities mentioned.

Going back to employment, having a physical or mental illness significantly decreases the chances of being employed. One study found that "thirty-two percent of clients ($N = 2{,}280$) reported they were unemployed during the entire year. Twenty-five percent reported that they were disabled" (Pickett-Schenk et al., 2002, p. 205). These startling statistics hold even 14 years later. It was found that those who reported severe depression worked part-time on average 1% in the previous year and those with a physical disability worked 1% of the time compared to the rest of the population. When it came to those who were unemployed and not looking, it was found that 11.2% had physical disability, 6.3% had severe depression, PTSD followed at 4.2%, traumatic brain injuries at 1.8%, and developmental disability at 1.1%. It must be noted, to avoid any confusion, that the aforementioned statistics are compositions of the entire raw population. The reason unemployment and not looking category exhibited such high levels of unemployment is due to the incapacitating nature of the conditions present. The following section concerns government assistance use.

Government assistance consists of a wide array of programs any American citizen can enroll in during times of need or assistance. The government (social services) has the ability to offer housing and food vouchers, immigrants benefits, it can offer Medicaid or Medicare, dispenses social security for disability or retirement, supplies veterans with much needed care, etc. The types of government assistance programs that were listed are as follows with the corresponding population percentage: CalWORKs (1.2%), CAPI (.5%), Child Support (.7%), SCHIP (.2%), Food Stamps (31.3%), GRGA (General Relief or General Assistance) (31.2%), HealthWayLA (1.5%), Medicaid (14.1%), Medicare (1.8%), Social Security Disability (17.1%), Social Security Retirement (1.8%), Unemployment (1.1%), Veteran Disability (.7%), VA Benefits (.7%), Veteran Pensions (.5%), WIC (.2%), and finally None (32.3%). Throughout this list, there were a couple of notable discrepancies that must be addressed.

First, the three things that stand out in this category are the high levels of General Relief or General Assistance, EBT, and Medicaid/Medicare. GRGA offers benefits and reliefs through the form of financial aid, while food stamps provides food vouchers for nutritional products. Finally, Medicare/Medicaid alleviates medical costs when required. Thankfully enough, these programs offer quite a bit of sustenance to the homeless population, for example, "such institutions offer hunger-targeted aid, such as the federal government via food stamps and the many local organization that run soup kitchens, mobile food unites, and food pantries" (Lee & Greif, 2008, p. 7). This is not a new concept and will (hopefully) be continued to be heavily funded. But, although food stamps are readily accessible, many chose to not pursue the option due to being barred or outright rejecting the system.

For example, one study notes that "not all homeless people take advantage of the infrastructure, given eligibility problems, lack of access, or a desire to minimalize dependence" (Lee & Greif, 2008, p. 7) The LAHSA survey found that 32.3% of the population did not utilize any form of government assistance program. This may be attributed to stigma associated with accepting help from the government. In a way, rejecting the government assistance programs salvages the image of the self by conceptualizing self-reliance. Another study found that, "they (the homeless) did not feel that they were treated as fully recognized adults or respected as equal citizens, but rather as members and children" (Hoffman & Coffey, 2008, p. 205). Certain portions of the homeless population would eventually forgo social services due to a lack in preferential treatment as fully grown American citizens. The homeless believe they are entitled to service with regards to their sentiments and vast array of situational characteristics. A client reported that "you walk into a food stamp office; I feel that I should be cared about. Show me concern, walk the path … with me. If you burn out from your job, move on. Give it to somebody else that cares … but don't just sit there and take it out on me" (Hoffman & Coffey, 2008, p. 215). Instead of providing consideration of the services provided, the social workers look down and belittle the population. It must also be noted that barriers to food stamps is typically due to having a criminal record. Barriers to entry and stigma eventually lead to the non-utilization of these services.

The following section is devoted to the population that utilizes Medicaid/Medicare. The homeless population are exposed to a variety of health maladies that require immediate medical attention and a large portion of the population will typically forgo this "luxury" in order to survive another day. Money is scarce, but poor lifestyles, drugs use, sexual abuse, physical, and mental maladies should never inhibit any individual from being admitted to a medical facility or utilizing its services. LAHSA reports that 15.9% of the population has either Medicaid or Medicare, and yet 25–30% of the population have been to the ER in the past year. In addition to this finding, it must be noted that the current study did not ask the population of their current insurance predicament. Medicare/Medicaid use cannot be emphasized enough to sustain the population, especially those who are homeless with severe mental illness. A study notes that this population "suffers more severe psychiatric symptomology, have higher rates of inpatient admissions, and are more likely to abuse alcohol and drugs" (Pickett-Schenk et al., 2002, p. 200). The high rates of mental illness on the street further exacerbates the needs of the disenfranchised in LA County.

So, what is the gender composition of the population? The study found that 72.5% of the population identified as male, which is typical finding, but women composed of 25.1% of the population. This number seems remarkable high as women are an unseen population throughout LA County; in all actuality, the LA Times reported that women are almost a third of the entire homeless population in Los Angeles County (Schluess & Elebee, 2016). It is interesting to note that gender is steady in relation to being homeless and a veteran. For example, "veteran status and black race significantly increased the risk for homelessness for both men and women" (Fargo, 2012, p. 1). In the following sections, the topic under discussion will be gender in association to physical/sexual abuse and dating violence or stalking.

Of the entire raw population, 17% of the homeless population had experienced dating violence or stalking in their lifetime. Those who had experienced physical or sexual abuse correlates with close to a fifth experiencing the negative adversity. This was an interesting find as "one study indicated that up to 50% of homeless and runaway adolescents may have experienced physical abuse; almost one-third experienced sexual abuse" (Shelton, Taylor, Bonner, & van den Bree, 2009, p. 465). While the LAHSA study produced results far below the 50% mark, 19% of the population being physically or sexually abused is highly significant. In terms of gender, men faced greater physical or sexual abuse than their female counterparts, 51% versus 47%. Another study found that "women may also be more likely than men to experience violence and sexual victimization during homelessness" (Meanwell, 2012, p. 73). A solution to this gender disparity may result from the confusion of either the physical/sexual abuse question. To resolve this issue, it has been suggested to the researchers to divide this question into a binary one. Men may have faced more physical abuse, while women may face a significant share of physical/sexual abuse while being homeless.

Statistically speaking, the homeless are bound to have an encounter with criminality within their lifetime, therefore this section regards criminal attributes. A study found that spending time at a foster care, being in juvenile detention, or in jail or prison, or being an adult probationer, or being on parole had potential higher risk of being homeless compared to those who had never been institutionalized (Shelton et al., 2009). The LAHSA study found that 10% of the population had been in foster care at least in one point in their lives. Shelton even discovered that "rates of adoption and fostering placement were also higher than on the never homeless group" (Shelton et al., 2009, p. 468). The study concluded that those who have never been adopted, faced a 1.8% possibility of being homeless and those who had been adopted saw the number dramatically rise to 12%. The same can be said of those who have been to juvenile detention or a probation center. 13% of the LAHSA population reported ever being send to the aforementioned institutions. Shelton hypothesizes that "a temporal relationship between crime and homelessness, it is interesting that a recent study of 14–25 year olds found that criminal behavior followed, rather than preceded, homelessness" (Shelton et al., 2009, p. 471). Ultimately, he concludes that there is a trend between criminality and adolescents with being homeless. Ultimately, certain factors increase the possibility of being homeless and preventive measures should be set in place to reinforce the individual's actions and lifestyle.

LAHSA discovered a startling statistic that 51% of the homeless population in LA County have been incarcerated in either jail or prison. While this statistic is strangely high, Fischer notes that a majority of the crimes are property ones. For example, he notes that "burglary charges resulted when the homeless were discovered sleeping in vacant buildings (and) in 28% of these reports, there was clear indications that the suspect was mentally disturbed or intoxicated" (Fischer, 1992, p. 59). One study went as far to report on the bureaucratic victimization of the homeless in effort to criminalize the population and protect the non-homeless ones. The Safer City Initiative in Los Angeles was meant to prevent criminality in Downtown Los Angeles by dismantling their homes due to health code reasons, asking the homeless population to relocate (forcing them to leave their belongings), and by incarcerating them for being homeless. The full aim of the initiative was to "reduce nuisance, property, and violent crimes that were thought to be associated with the spatial concentration of homelessness in Skid Row" (Culhane, 2010, p. 851). Therefore, instead of assisting the homeless in their condition, the city further criminalized the population for being homeless which proves to be counter-productive, as being arrested limits the opportunities a homeless individual can achieve. In recent light, it is fair to note that "criminal activity among the homeless exceeds that of the general population" (Fischer, 1992, p. 60).

Even though 50% of the population had been incarcerated, the data found that 27% of the population has been on adult probation and 20% of the population have been on parole. Unfortunately, no age statistics were available to know when the population had been criminalized. With the data at my disposal, I decided to compare employment opportunities with jail time. It was found that less than 4% of the entire population had been incarcerated and held a full or part-time position. This is significant given the high percentage of the incarceration rate. On the other hand, close to 30% of the homeless population had been incarcerated and had no employment and were not actively looking. The nature of this paradigm may derive due to the stigma an individual may face and the barrier to employment entry due to being homeless and previously being incarcerated. For example, Wenzel notes that stigma leads to "Unemployment and repeated failures to procure decent jobs may contribute to a lack of hope, diminished self-confidence, reduced commitment and even eventual withdrawal from extensive employment activities" (Wenzel, 1992, p. 57). The problem of hopelessness and unemployment among the homeless community must be reevaluated to establish an association for positive employment opportunities, which in turn would equate to reintegration.

The second to final section of this essay concerns where the homeless come from and how they utilize their space. In terms of space utilization, it was found that the homeless mostly slept on the street, sidewalk, or an alley with a strong composition of 53% of the total population, while 9% slept at a park, beach riverbed, and only a meager 6.6% slept in a tent. These harrowing statistics exemplify the true nature of a homeless person's experience, considering most sleep without adequate shelter or clothing. The other 30% rested in a variety of other urban sprawling's such as campgrounds/woods (1.3%), under bridge or overpass (3.7%), other outdoor location (2.8%), abandoned building or parking lot (3.1%), car or truck (7.1%), RV or camper (4.7%), unconverted garage, attic, or basement (1.4%), and finally other makeshift shelter not meant for human habitation (2.6%). Those who were able to procure shelter would be considered under the car or truck and RV category, which boasts more individuals than a tent.

The LAHSA data did find that the majority of individuals lived in LA County prior to being homeless. LA County residents compose of 57%, while the remaining portions of California compose a measly 11% and those who are from out-of-state composited 15% of the population. The reason LA County may be such a hot bed for homelessness is the fact that the rent is high, translating into residents not being able to afford rent, and the supportable tropical climate. Although the County boasts a wonderful climate, residents have been let down by Los Angeles County. It is disturbing to find a majority of residents are from the same populace.

Finally, this study moves onto veterans. Surprisingly, veterans compose of a small portion of the entire population as they surmise a little over 5% of the sampling frame. A majority of veterans have had an honorable discharge, but

90% of them are unemployed. This is an underserved community, but this finding should correlate with VA benefits or other forms of government assistance. Unfortunately, less than 1% have Veterans Disability or VA Benefits or A Veteran Penion's claimed.

CONCLUSION

There is so much data left to be collected and analyzed, that I cannot even fathom processing at the moment. LA County residents are sleeping on the stress of downtown and in other unreasonable areas. Homeless individuals deserve the same respect and opportunity for success even when times are down, which may force them into criminal behavior or other forms of delinquency. For every act there is an effect, but reprimanding the population for being poor, being physically or mentally unable, or being a criminal should not bar an individual from employment or government access. By doing so, Los Angeles County only perpetuates the system of homelessness and obstructs a large population from being contributive members of society. The population needs to know that the people are on their side, and constructing affordable housing is a great place to start, but it is not enough. The homeless need to have access to health care, nutritional sustenance, employment, and a roof over their heads to begin anew. By doing so, not only does the homeless population has a new place to begin, so do we.

LAHSA has been a primary contributor on this project and the questions posed on government assistance and employment were additions to their already year by year study. To cement the internal validity of the data, the following year (2017) there should be renowned interest on the topic, especially in this subsect of demography. Ultimately, I hope this project will bring insight into a homeless individual's experience, for their existence is being cataloged. This chapter has been made accessible to the public for anyone to read, for this is my role as a public sociologist.

ACKNOWLEDGMENT

I would like to thank Lorin Kenny for his assistance on this project and the data.

REFERENCES

Agans, R. P., Jefferson, M. T., Bowling, J. M., Zeng, D., Yang, J., & Silverbush, M. (2014). Enumerating the hidden homeless: Strategies to estimate the homeless gone missing from a point-in-time count. *Journal of Official Statistics*, *30*(2), 215–229.

Borchard, K. (2005). *The word on the street: Homeless men in Las Vegas*. Reno: University of Nevada Press.

Culhane, D. P. (2010). Tackling homelessness in Los Angeles' Skid Row: The role of policing strategies and the spatial deconcentration of homelessness. *Criminology and Public Policy*, *9*(4).

Fargo, J. (2012). Prevalence and risk of homelessness among US veterans. *Preventing Chronic Disease*, *9*.

Fischer, P. J. (1992). The criminalization of homelessness. In *Homelessness* (pp. 57–64). New York, NY: Springer.

Hoffman, L., & Coffey, B. (2008). Dignity and indignation: How people experiencing homelessness view services and providers. *The Social Science Journal*, *45*(2), 207–222.

LAHSA (2016). *The Los Angeles Homeless Services Authority Continuum of Care Homeless Inventory Count (LAHSA) 2016*, Los Angeles Homeless Service Authority.

LAHSA (2017). *The Los Angeles Homeless Services Authority Continuum of Care Homeless Inventory Count (LAHSA) 2017*, Los Angeles Homeless Service Authority.

Lee, B. A., & Greif, M. J. (2008). Homelessness and hunger. *Journal of Health and Social Behavior*, *49*(1), 3–19.

Meanwell, E. (2012). Experiencing homelessness: A review of recent literature. *Sociology Compass*, *6*(1), 72–85.

Milby, J. B., Schumacher, J. E., Wallace, D., Vuchinich, R., Mennemeyer, S. T., & Kertesz, S. G. (2010). Effects of sustained abstinence among treated substance-abusing homeless persons on housing and employment. *American Journal of Public Health*, *100*(5), 913–918.

Pickett-Schenk, S. A., Cook, J. A., Grey, D., Banghart, M., Rosenheck, R. A., & Randolph, F. (2002). Employment histories of homeless persons with mental illness. *Community Mental Health Journal*, *38*(3), 199–211.

Poremski, D., Whitley, R., & Latimer, E. (2014). Barriers to obtaining employment for people with severe mental illness experiencing homelessness. *Journal of Mental Health*, *23*(4), 181–185.

Schluess, J., & Elebee, L. (2016, October 28). Retrieved from http://www.latimes.com/projects/la-me-homeless-los-angeles-2016/. Accessed on November 27, 2016.

Shelton, K. H., Taylor, P. J., Bonner, A., & van den Bree, M. (2009). Risk factors for homelessness: Evidence from a population-based study. *Psychiatric Services*, *60*(4), 465–472.

Wenzel, S. L. (1992). Length of time spent homeless: Implications for employment of homeless persons. *Journal of Community Psychology*, *20*(1), 57–71.

Wesely, J. K. (2009). "Mom said we had a money maker": Sexualization and survival contexts among homeless women. *Symbolic Interaction*, *32*(2), 91–105.

POLICE CULTURE, GENDER AND CRIME IN THE IRISH FREE STATE

Anthony Keating

ABSTRACT

Sexual crime in the Irish Free State was more than an issue of law, it carried ideological importance in a nation that legitimised itself as a beacon of Celtic Catholicism whilst struggling to maintain credibility in a contested post-colonial landscape. The nation's police force, An Garda Síochána, had a central role in preserving the nation's reputation for piety. This chapter explores the views of two of An Garda Síochána's most senior officers regarding female sexuality and sexual crime; features that were to influence the level of protection and justice Ireland's women and children were afforded under law.

Keywords: Police culture; gender; Irish Free State

INTRODUCTION

Sexual crime in the Irish Free State[1] was an issue that, whilst carefully managed and downplayed in the public arena, caused alarm amongst the country's political and religious elites (Keating, 2012b). Sexual immorality and crime were regarded as issues that threatened to undermine a central myth at the core of the fledgling nation's post-colonial identity, namely, that the Free State was a beacon of Catholic Celtic purity in a world otherwise corrupted by vice

Environmental Criminology: Spatial Analysis and Regional Issues
Advances in Sustainability and Environmental Justice, Volume 20, 87–101

ISSN: 2051-5030/doi:10.1108/S2051-503020170000020006

(Keating, 2012a). Sexual crime was not only viewed as an issue of personal tragedy and law enforcement but also viewed as an issue that had the potential to undermine the religio/national image so important to the emerging state, on both the national and international stage.

This chapter will explore the views of the Free State's most senior police officer O'Duffy and his Deputy Commissioner relating to immorality and sexual crime in the opening decade of the Irish Free State, the era in which many of the mores embed in Irish culture in the pre-revolutionary period were reinforced and strengthened in Irish life (Inglis, 1998; Keating, 2012b). The views they expressed will be contextualised within the wider ideological, religious legal and social attitudes in the ascendency in the Free State. It will be argued that the attitudes articulated by these senior police officers relating to female sexuality, immorality and sexual crime resulted in women and children having a heightened vulnerability to sexual crime with victims and their families far less likely to report sexual crime, in part these victims feared that reporting sexual crimes against them could result in them being punished and stigmatised far more harshly than those who assaulted them, even in the unlikely event of their attackers conviction.

The nation's police force, An Garda Síochána, was envisioned by its founding Commissioner,[2] Garda Commissioner Eoin O'Duffy, as playing a central role in the shaping of the nation's Catholic social order, as well as policing its laws (Finnegan, 2004). O'Duffy took a particular interest in sexual crime and prior to giving evidence to the Carrigan Committee[3] in 1930, one of his two Deputy Commissioners, either W.R.E. Murphy[4] or Eamon Coogan,[5] drew up a report on sexual morality in the Free State which mused over the causes of and remedies for sexual crime. The document ends with the words 'Leas-choimisinéir', the Irish for 'Deputy Commissioner' but unfortunately no individual's signature or other indication of which of the two Deputy Commissioners authored the document. This document, when combined with O'Duffy's evidence to the Carrigan Committee, provides insight into the attitudes towards sexual crime at the top of Irish police's hierarchy. O'Duffy and his deputies played a key role in setting the culture of An Garda Síochána consequently contributing to the reception that complainants and alleged perpetrators would receive from the police, thereby influencing the propensity of those who have been the victims of sexual crime to seek justice under the law (Bourke, 2008). Their views on this issue created a cultural paradigm that would endure, with greater or lesser degree of subtlety, into modernity.

Mary McAleese, the then President of Ireland speaking at the 30th anniversary of the establishment of the Dublin Rape Crisis Centre in 2010, commented upon the culture of non-reporting or under-reporting of sexual crimes fuelled by fear and a lack of faith in the policing and criminal justice system amongst victims, which, she argued, created a 'rapists' charter' that gives sexual offenders 'free rein'. The President highlighted the role victim's sexual histories still played in discrediting her as a complainant, both in the eyes of juries and the wider community, a feature she described as a 'still a neuralgic and fraught' issue in sexual assault cases in contemporary Ireland (McDonald, 2010).

James Hamilton, the then Director of Public Prosecutions, speaking at the same conference, acknowledged that there was a 'huge problem' relating to attitudes to sexual crime in Ireland. A problem Hamilton argued is particularly prevalent in small communities, asserting 'The criminal justice system cannot change the attitudes of neighbours or people in small towns' (*ibid.*). The 'attitudes' Hamilton was referring to were clearly evident in two relatively recent rape convictions in Ireland, namely that of a Danny Foley in Listowel, Co Kerry and Liam Sheehy in Daugh, Co Kerry.[6] The attitudes displayed in these cases towards the victims of sexual crime by sections of the community have a long-standing history in Ireland, whilst they can in no way be described as uniquely Irish features (Bourke, 2008) they have been particularly persistent in the Irish context as a result of Ireland's unique socio-religious history (Keating, 2012a). Irish policing and the wider legal system has proved itself adept at failing to protect vulnerable sections of society from sexual crime and the moral approbation of a citizenry heavily influenced by puritanical sexual mores (Finnegan, 2004; Keating, 2012a; Ryan, 2002).

SEX AND MORALITY IN THE IRISH FREE STATE

Sexual crime in the Irish Free State was viewed and managed through a lens created, in part, by the ideological construct born of an imagined past Irish nation of Celtic-Catholic purity reborn in post-colonial/post-revolutionary Ireland (Keating, 2002; Larmour, 1998; Ryan, 2002; Szmigiero, 2006) and by a fearful pessimism relating to human sexuality that pervaded the Irish elites and 'respectable' Irish society. Evidence exists that suggests that this view of human sexuality was not universally shared by all sections of Irish society[7] and that the elites, who represented and reflected what it was to be respectable and truly Irish (Keating, 2012a), were determined to manage the danger they perceived in human sexuality by law and cultural and economic pressure (Keating, 2012a).

The Irish revolution was male, Catholic and petty bourgeois, shaped, as Brown has observed:

> By 'the farmers and the tradesmen together with such of their offspring as could find roles in the professions' who were 'enormously influential in fashioning the political, social and cultural molds of the independent state. Their economic prudence, their necessarily puritanical, repressive sexual mores and nationalistic conservatism, encouraged by a priesthood and hierarchy drawn considerably from their number, largely determined the kind of country which emerged in the first decades of independence'. (Brown, 2010, p. 23)

Free State Ireland was an imagined, as well as an actual, nation, deeply rooted in the values of an orthodox Catholic, petty bourgeois, male elite, deeply fearful of sex and sexuality, particularly that of female sexuality (Ryan, 2010). A fear born from the particular form of authoritarian Catholicism referred to by Brown, as 'Jansenistic Puritanism'[8] (Brown, 2010) and a nationalist 'Victorianism' that sought to emulate the imagined purity of an earlier pre colonisation Celtic age, which, conversely, lead to Irish women being restricted by law in terms of their sexual rights in ways unknown under ancient Irish Brehon laws[9] (Ryan, 2010). The new State, not only embraced laws established by the British that disadvantaged women, but added to them in the shape of laws that censored debate and expression, control of fertility, excluded women from employment and protection under the law[10] (Ryan, 2010).

Idealised representations of Irish women and manhood in ideological nationalism played a significant role in the framing of, and response to, sexual crime in the Free State. (Ryan, 2002). Given the symbolic representation of Irish womanhood as the Irish nation, raped by the colonial oppressor and the heroic role of Irish men in freeing female-gendered Ireland from her colonial rapist oppressor, the post-colonial government found it undesirable on an ideological level for Irish men to be seen to be engaged in this form of abuse, or indeed, sexual immorality in general.

The State was so concerned that its troops should not be associated with immoral activity that it went as far as to suppress the 1926 Report of the Committee on Venereal Disease. This Report, rather shockingly, from the State's perspective, provided evidence that the spread of sexually transmitted disease was a phenomenon that correlated with the movement of Free State troops around the country undermining the cherished view that it could be identified exclusively as a result of urban prostitution (Howell, 2003). The revelations regarding venereal disease overturned a long-held view, which provided some comfort to the religious and political elite, namely, that 'country districts and Ireland as a whole were largely blameless; contagion and immorality could be traced fairly and squarely to the large city; to the corrupting effect of colonial occupation' (Howell, 2003, p. 326).

The virgin and the married mother were the two towering and acceptable stereotypes of Irish womanhood drawn from Irish Catholicism's particularly strident Marian devotion[11] (Beaumont, 1997). Irish women were left in no doubt that 'the Irish woman who is ashamed to be a true child of Mary is a sham Irish woman'[12] (Beaumont, 1997, p. 567). Irish women in the Free State were charged with prime responsibility for national virtue and morality. The sexual purity of the nation's women provided the underpinning that legitimised their role as cultural custodians of the true Irish virtue assigned to them by the ideologues of Catholic nationalism (Valiulis, 1995a). Irish manhood's role in ensuring Ireland's bloodline and cultural legacy was in policing his own 'baser' instincts and protecting Irish women from predatory males and their own innate feminine weakness. Irish women were, for all their cultural importance, viewed by the Free State's religio-political elites as 'fragile, weak beings', beings dependent on 'virtue and chastity' in order to maintain their 'moral power' (Inglis, 1998, p. 210).

The Free State assigned to Irish women 'the responsibility of ensuring the "Irishness" of the new state, that is, of preserving and transmitting traditional Irish culture' (Inglis, 1998, p. 129), a cultural ideal in which sexual immorality had no place. Therefore, if sexual immorality was abroad amongst sections of womanhood in the Free State it followed that its existence must be the result of something other, something alien. The religious and political elite of the Free State resolved these difficulties by externalising the potential threat. The 'flapper'[13] became the actual and metaphorical embodiment of female corruption in the Irish Free State (Keating, 2012a; Luddy, 2007; Ryan, 2002). She along with the ill-disciplined, only partly formed adolescent personality, with its proclivity for corruption, became the objects of fear and control by Catholic and Free State institutions. A fear which found its praxis in rigid social controls that utilised the moral authority of the Catholic Church and the legislative power of the State to introduce laws relating to contraception, censorship, dance halls and female employment (Horgan, 1995; O'Conner, 2003; Valiulis, 1995a, 1995b).

These laws represented an attempt at controlling female independence, sexually and financially, and to curtail the vehicle for the introduction of these foreign vices into Ireland; vices, it was argued by the Free State's religious and political elites, that could destroy the Free State project before it had a chance to emerge fully formed from the innate virtue of the Irish people guided by the Catholic Church (Keating, 2012a).

The evils of sexual promiscuity and sexual crime were, at least for public consumption, external threats the product of foreign corruption or the result of aberrations of developmental or mental instability (Keating, 2012a). During the post-revolutionary Free State, a motif of the alien was developed to explain the actions of those, male and female, who dissented from the prevailing moral code (Keating, 2012b; Ryan, 2002). Women corrupted by this virus were not the 'true children of Mary' they were dangerous 'moderns', hell bent on their own destruction and the destruction of the fledgling Free State. The prostitute, it was argued, was the obvious, visible tip of the danger, whilst the modern young women harboured the real potential for corruption. The dangers they presented were more insidious and occult than those of the 'fallen woman' driven to vice unwillingly by scheming men and economic necessity.

Clerics, the majority of Irish politicians and a substantial swathe of 'respectable' public opinion, believed that any deviation from the strictly defined 'legitimate' sexual activity of a woman, even against her will, was a woman's responsibility and she had therefore failed or indeed undermined her sex and race (Valiulis, 1995b). The concept of 'ruin' was a very real one for Irish women in this period (Larmour, 1998; Ryan, 2000). One of the possible consequences of transgressions against the moral code, even unwillingly, could be incarceration in a Magdalen Asylum[14] (Ferriter, 2005; Finnegan, 2004).

In addition to any reputational constraints, the law relating to the gathering of evidence and the reporting of sexual cases provided another inhibitor to women reporting or pursuing justice (Keating, 2012a). When an allegation was made, the first stage of the process was the preparation of a deposition. This involved a detailed recording of the allegation by a police officer; however, the alleged perpetrators had the right to be at the deposition and to personally, or through their lawyer, cross-examine the deponent. The surviving archival material shows clearly that alleged perpetrators could make full use of this opportunity to undermine and intimidate their accusers (Various, National Archives of Ireland (NAI), Department of Justice (DJ). H234 series).

In addition to this, if the matter preceded to trial, a number of factors impacted on a woman's ability to receive, or at least be confident in receiving, justice and protecting her reputation (Beaumont, 1997). This imbalance was compounded by a legal rule of practice which existed that required a judge in cases of sexual offences, treason or perjury to warn the jury of the risks inherent in convicting on the evidence of a single witness (Larmour, 1998).

There was also no protection for the alleged victims during the trial in regard to publicity. Where coverage was given to a trial the name, address and other personal details of the alleged victims could be printed with no consideration of age or any other sensitivity. The threat of this public exposure was real and although far from guaranteed[15] the potential public exposure and the serious consequences that this could have on a woman's life chances added a further burden on women seeking protection and redress for sexual assault under the law.

THE DEPUTY COMMISSIONER'S PAPER

The document, Garda Archive S.D.323, dated 28 March 1929, begins with a long, rather Hobbesian,[16] preamble on the nature of the social order and male sexuality. The Deputy Commissioner was far from being alone amongst the country's elites, civil servants and politicians, who along with the Catholic clergy were prophesying the imminent collapse of the social order if radical action was not taken to halt the nation's slide into moral degeneracy (McGarry, 2005; White, 1971), this theme, in many ways, was a fixation for cultural-nationalists still fighting the old enemy, England, by battling its most pervasive legacy, sexual immorality.

Whilst it is impossible to confirm that the Deputy Commissioner's document was prepared for O'Duffy's evidence session for the Carrigan Committee, its date and the subject matter makes it highly likely that it was; therefore, it is, an important document at a critical juncture of the development of the Irish Criminal Code regarding sexual crime (Finnane, 2001; Keating, 2012b; Kennedy, 2000).

The preamble to the report is worth quoting at length as it carries a clear articulation of Murphy's views on the social significance of religious order and the nature and management of sexual deviance in Ireland prior to the inception of the Free State.

The document is unequivocal in the view that societal order in general was based to a very large degree on the close control of sexual morality. Indeed going as far as to state that those who opposed the existing social order did so in general as a result of their sexually 'moral depravity'.

The Deputy Commissioner asserts repeatedly that males are driven by primeval sexual drives and that modern, socially constructed, sexual convention was 'an unwelcome compromise between the individual man's sex needs and the needs of the community to which he belongs ...'. He develops his argument, asserting that civil strife can lead to a breakdown in this 'unwelcome compromise' that cause men to 'seek the satisfaction of his lower and thwarted appetites on his woman slaves'. He seems to be indicating here that prior to the First World War, the level of sexual promiscuity was much lower than contemporary views amongst the police suggested was now the case. The lack of available evidence makes it extremely difficult, if not impossible, to ascertain the true extent of sexual conservatism and adherence to church teachings, in this regard, prior to the First World War.

The anthropological work of Arensburg and Kimball in rural Ireland in the 1930s[17] cast some doubt on the longevity of puritanical sexual attitudes in Ireland (2001). Extrapolating from their work, Larmour has suggested that these attitudes were more liberal than Murphy claimed in his briefing document. She suggests that in the early years of the Free State the law was used to '... impose Roman Catholic middle class socio-moral values on sections of the community which had not yet adopted them' (Larmour, 1998, p. 3). There are however figures that indicate prosecution rates for sexual crime rose in Ireland by 25% in the pre to post World War 1 period (Finnane, 2001).

The Deputy Commissioner was clear as to where this perceived state of affairs had found its genesis, namely, in the 'pro-British women of Ireland' who 'made heroes of their soldiers and ... fools of themselves' as the soldiers, due to the morally corrupting influence of the war, 'reverted to the passions and appetites of the primitive and demand created supply'.[18] He asserted that this 'corruption' had infected the greater swath of Irish womanhood who, he argued, had in many cases succumbed 'to the appetites of their pro-British sisters', despite the 'new protection for its virtues' afforded to the Irish Race 'in the idealism of the men of 1916'.

The role of foreign influence over Ireland's young women was a major concern for the Deputy Commissioner. He quoted the Bishop of Anchory in his report, who observed that 'the refined modesty so long distinctive of our nation is already robed in cere-cloths and only awaits the pall-bearers of further foreign innovation to carry it to the grave'. He quotes another Catholic clergyman 'of experience and restraint' who bemoaned the alien influences on Irish youth asserting that 'one might spend days throwing stones into a crowd of 500 young men-or young women ... without any risk of hitting a clean living one'.

Irish women, for the Deputy Commissioner, or rather a certain type of Irish woman, despite the self-sacrifice of Irish men, were responsible for the breakdown of order in regard to sexual morality which in turn left them exposed to higher levels of sexual crime. He argued:

> The modern woman acts as though the destruction of the onetime established standards of morality were for her a highly desirable object, as if indeed the height of her ambitions were to smash the social system, which, with the consent of man, its principle victim, ensured, by marriage and family life, that her part in the reproduction of the species was if not easy, at least well protected, dignified and as satisfactory as it could be from her point of view.

He went on to develop this theme, asserting that the women of Ireland were acting out a 'suicidal folly' which was 'attributable in large measure' to the 'the present collapse of morals and further degradation and ruin which awaits us if not quickly averted'.

The Deputy Commissioner concluded his briefing document on rather a pessimistic note. He feared that the woman of Ireland could not be redeemed to their pre-1914 levels of virtue. He argued that the Catholic Church had become ineffective in terms of influence over modern young women and that neither the Church nor State should be openly associated with declining morality, for ideological reasons, although he felt the State and Church should ensure the road was kept open for those fighting the good fight against sexual immorality and crime. He concluded by suggesting, 'Perhaps a woman's movement would be best designed to succeed. After all it's a woman's mission and it's woman's interests that are at stake'.

O'DUFFY'S EVIDENCE TO CARRIGAN

O'Duffy's evidence to the Committee was broad-based, covering issues relating to the age of sexual consent, prostitution, brothel keeping, public indecency, homosexuality and most notably, issues relating to the sexual abuse of children, which have been covered in detail elsewhere (Finnane, 2001; Keating, 2002). O'Duffy's evidence was based on a detailed survey compiled by the Superintendents and Sergeants of over 800 police districts that covered the State and led the Committee to conclude that 'the moral outlook of the country had changed for the worst in recent years ...' (NAI. DJ: H122). O'Duffy asserted:

> Immorality in the individual or community is undoubtedly the forerunner of crime, and the latter can invariably be described as the effect of the former. In this country a careful examination of our most heinous crimes reveals the fact that in the majority of cases the depraved career of the perpetrator originated from his immoral instincts, or from association with persons of this class. The outlook of many of our people, even in rural areas, has changed within the past ten years, and the morally depraved who then would be exorcised from society are now regarded as rather clever and interesting. (p. 2)

O'Duffy was a complex man, with a deeply held Catholic faith, and profoundly right-wing in his politics, he would later lead Ireland's right-wing Blue Shirts and latterly the more clearly Fascist, National Corporate Party.[19] O'Duffy was something of a fantasist with a propensity for megalomania[20] who had attempted to develop a 'cult of personality' around his leadership of the fledgling Garda Síochána (McGarry, 2005). There is also evidence to suggest that O'Duffy struggled with his own, homosexual, identity (McGarry, 2005), something which did not stop him, or indeed may have stirred him, to be aggressively homophobic in his public utterances. O'Duffy was relentless in his public pronouncements regarding morality and continually cast the Garda as the guardians of Ireland's moral values.

The evidence O'Duffy presented, regarding prostitution, to the Carrigan Committee, shows him to be, by and large, paternally sympathetic to women engaged in prostitution, asserting, 'There is nothing inherently bad in the majority of our young Irish girls who drift into the unfortunate class' (NAI. DJ: H122.1). He viewed them as coming from the urban poorer classes, often abused as children or whilst working in domestic service. Many of these young women he asserted became unmarried mothers who were forced into the hands of unscrupulous degenerate men in order to survive, arguing:

> ... Generally the girls who resort to brothels and become common prostitutes are of respectable type who have given birth to illegitimate children and lost their employment. Through force of circumstance they drift into cheap lodging-houses where they become associated with prostitutes, and it is only a matter of time until they become common prostitutes themselves. These in turn encouraged by persons running 'Shebeens'[21] to entice males to these Shebeens to obtain intoxicating liquor, the sale of which at exorbitant prices appears to be the principle object of the owner or tenant of these premises.
>
> Facilities for immorality are usually afforded on the premises, or are available in the immediate vicinity (NAI. DJ: H122.4).

His focus was on removing the terrible social conditions, unemployment and overcrowded housing that he viewed as the causation of prostitution, the rehabilitation of those already engaged in prostitution, stricter punishments for men who used their services and brothel keepers. O'Duffy went as far as to suggest the removal of the assumption of innocence under law for those suspected of frequenting or keeping brothels and those he termed 'bullies', men who had sex with girls below the age of 13, the age of sexual consent at the time of his evidence being 16, arguing that 'the liberty of the citizen should not apply to such as these' (NAI. DJ: H122.12).

O'Duffy viewed the protection of teenage girls as a main priority, not just as an issue of the protection of children but as an issue with serious implications for the moral health of the nation, asserting:

> Between 13 and 18 character is formed – habits acquired then are assimilated and play an unmistakable part in the future conduct of the individual. Between those ages sexual passions are so predominant that they are most difficult to subdue, and females are affected with natural weakness. They have not the physical strength or strength of will to resist suggestions which so strongly appeal to their senses, nor have they the understanding to appreciate the consequences. They become an easy prey for those so inclined, and, having fallen at this impressionable age, there is little hope that their career can be anything but depraved. They

> will become a source of temptation to men, and their conduct will influence those of their sex with whom they come in contact. (NAI. DJ: H122.2)

O'Duffy's evidence to the Commission also illustrates an important nuance regarding his view of the dangers of female sexuality in general, a view that had important resonances for the Garda's approach to sexual crime. O'Duffy's view of women who engaged in sexual activity for their own gratification was far less generous than his attitude to that 'unfortunate class of girl' who had fallen into prostitution. He argued:

> While prostitution is not carried out in the rural parts, there are in every district women known as 'amatures' who are willing to gratify the passions of any man while not requiring to be paid for their services.
>
> Most of the women who allow men to carry on indecently with them in the public parks, and along the road-sides and lanes belong to this category and as such do more to corrupt public morals than the prostitute. The latter is known for what she is, and is avoided by right-thinking men and women. She is not causing social disorder or moral degradation in the same insidious manner as the so called amateur. (NAI. DJ: H122.17)

Female sexuality, in O'Duffy's view, was itself corrosive and far more dangerous than an industry populated by unfortunate 'fallen girls' giving themselves without pleasure for money, victims of poverty and aberrant male sexual desire. O'Duffy's views represented the mainstream of thought amongst Ireland's male elites, indeed his views relating to women working in prostitution may have seemed to some liberal, offering sympathy over blame and seeking to challenge male immorality and abuse of economic position. 'Fallen girls' were in some ways safe and not threatening, a historical apparition, visible and controllable, a group of women who things were done to rather than acting from their own sexual agency, in some way they were de-sexed in their representations, automatons rather than women.

'Amateurs', however, enjoyed sex for sex's sake, something that right-thinking Irish men and women should not do; in particular Irish womanhood, with its responsibility as a moral and cultural transmitter of the religion, race and culture. Irish manhood, had in the view of fundamentalist Catholic-nationalist ideologues, like O'Duffy, had sacrificed too much to achieve the freedom of mother Ireland and liberate Irish womanhood from the rapist coloniser too allow her to drag herself and race down as a result of her baser instincts. The 'amature' for O'Duffy was a form of 5th columnist poised to undermine faith and country. For O'Duffy and his Deputy Commissioner manifestations of corrosive femininity were everywhere, the sexual freedom indulged in by the 'amateur' was mirrored in calls for other freedoms. The right to wear revealing clothes, have a career, control fertility, dance to 'uncivilized' music were all manifestations of women's desire for their own downfall at a point in Irish history when Irish manhood had been corrupted and enfeebled by years of war, just the point at which they needed Irish womanhood to nurture them in safe respectable homes keeping them safe from temptation and their own 'base desires'.

O'Duffy was a tireless organiser with a clear vision. His men's training included clear prescriptions of what proper behaviour was and what a well-balanced moral society looked like. Sexuality and sexual immorality and crime were important parts of their leadership's agenda and a culture of suspicion of those who were found to be acting outside of right behaviour became embedded within Irish police culture. Any suspicion that women had opened themselves to the risk of assault by wearing 'immodest' clothing, their deportment or actions could lead women to being viewed as agent provocateur rather than a victim of sexual crime, which carried the risk of social alienation and even incarceration of victims to save society from the threat inherent in their sexuality.

CONCLUSIONS

Sexual crime in the Irish Free State became interwoven with the wider issue of sexual morality and this debate, in turn, was framed within an ideological discourse that viewed sexual morality as an integral part of 'real' Irish identity, an identity which viewed the advancement of the Catholic moral agenda as part of Ireland's historic mission. The sexual integrity of the individual Irish man and woman was an important building block of nationhood and any attempt, or perceived attempt, to undermine that morality was an undermining of the State itself. The Catholic-Nationalist mind set, viewed female sexuality and the modern woman's desire for freedom as a threat to the fledgling Free State's historic mission, above all they were not for public or international consumption, something that could spread the contagion at home and bring the Free State mission into question, particularly in the eyes of Britain, the ever-present, ever-threatening colonial power.

Women who contravened the behaviours expected of them, in the canon of religio-nationalist ideology, were viewed as agents of evil. In this ideological battleground, women who were victims of sexual crime could also fall under suspicion and in essence be punished for being sexually assaulted, through public humiliation, the loss of reputation and even incarceration in Magdalene Laundries. These were powerful inhibitors to women in both reporting sexual crimes and pursuing prosecutions. The structure of the criminal justice system and contemporary social norms reinforced these inhibitions and proved a powerful controller of women's behaviour, powerful enough to silence the ideologically inconvenient truth about rape and sexual assault in the Free State.

The Deputy Commissioners Report and O'Duffy's evidence to Carrigan reveal how fearful, that those at the top of Irish policing were of female sexuality and, what they believed to be, its potential to undermine the moral health of the Free State. The attitudes they portray were consistent with the wider Nationalist trope regarding women, and the Jansenistic Puritanism of Irish Catholicism.

Sexual crime in the Free State was not an issue for unregulated public consumption. It was a toxic subject handled by experts out of sight of public scrutiny, nationally and internationally. The real discussions on this subject were held behind closed doors. The Deputy Commissioners Report and O'Duffy's subsequent evidence to the Carrigan Committee offer one of the few opportunities to explore first-hand the views of leading actors on the issue. It demonstrates clearly that those at the very top of Irish policing viewed female sexuality itself, and the freedom being sought by 'modern' young women, as contributing to the level of sexual crime in the Free State and that women in general were, at least in part, responsible for the crimes committed against them. Women were viewed too often as agent provocateurs rather than victims of male sexual aggression. In some ways, women became the real enemy, the State had to control, an attitude that propagated a culture hostile to women and girls who were raped and sexually assaulted for decades to come.

NOTES

1. The Irish Free State (Irish: Saorstát Éireann) (6 December 1922–1937) was the state established as a dominion under the Anglo-Irish Treaty, signed by the British government and Irish representatives. The Irish Free State came to an end in 1937, when the citizens voted by referendum to replace the 1922 constitution. It was succeeded by the sovereign, modern state of Ireland.

2. The original police force of the Irish Free State, the Civic Guard was formed by the Provisional Government in February 1922 to take over the responsibility of policing the fledgling Irish Free State. It replaced the Royal Irish Constabulary (RIC) and the Irish Republican Police of 1919–1922, commanded by Michael Staines who only remained in post for 8, resigning in the September of 1922. O'Duffy became Commissioner of the Civil Guard on Staines's resignation, overseeing its change of name under the Garda Síochána (Temporary Provisions) Act 1923. O'Duffy, who remained in post until 1932 is widely credited as the founding leader of An Garda Síochána, the man who moulded them into a disciplined professional force, established its disciplinary codes and working practices (McNiff, 1997).

3. In 1930 the Minister for Justice, James Fitzgerald-Kenney established the Committee on the Criminal Law Amendment Acts and Juvenile Prostitution under the Chairmanship of William Carrigan QC. The Carrigan Report was presented to government in 1931 and played a significant role in the creation of the Criminal Law Amendment Act (1935). However, the report was suppressed from public consumption due to the shocking revelations relating to sexual crime in the Free State. The report was not available for public scrutiny until its release to the National Archives in 2000.

4. William Richard English-Murphy DSO MC MiD known as W.R.E. Murphy (1890–1975) was an Irish soldier and policeman. He served as an officer with the British Army in the First World War and later in the Irish Army in the Irish Civil War. In the Civil War he was second in overall command of the National Army from January to May 1923. He was the last Commissioner of the Dublin Metropolitan Police before its merger with the Garda Siochána in 1925. Thereafter he was the Deputy Commissioner of the Gardaí until his retirement in 1955 (Murphy, 2005).

5. Eamonn Coogan (1896–22 January 1948) was an Irish Fine Gael politician, barrister and Deputy Commissioner of the Garda Síochána. He was appointed to his role as Deputy Commissioner whilst serving as a civil servant in the Department of Local Government.

6. At Foley's sentencing hearing, up to 50 people, mostly middle-aged men, queued to sympathized with him and the victim spoke of being 'judged' by the community for

taking him to court. Similarly in the Sheehy case leading members of the community sought to defend his reputation and dozens of locals stood to publicly shake his hand at his sentencing hearing.

7. Arensberg and Kimbal's Family and Community in Ireland, an anthropological study conducted in Co Claire in the 1930s, described patterns of sexual attitudes that did not fit the Free States view of sexual propriety. Another important text in gauging the realities of sexual life in the Irish Free State is The Tailor and Ansty by Eric Cross published in 1942 book about the life of the Irish tailor and storyteller, Timothy Buckley, and his wife Anastasia ('Ansty'). The book was banned by the Censorship of Publications Board because of its depiction of premarital cohabitation, and its sexual frankness (Cross, 1981).

8. Jansenism was a Catholic theological movement, primarily in France, that emphasised original sin, human depravity, the necessity of divine grace and predestination (Brown, 2010, p. 23).

9. Early Irish law, sometimes called Brehon law, comprised the statutes which governed everyday life in Early Medieval Ireland.

Under the Brehon Laws women had greater freedom, independence and rights to property than in other European societies of the time. Men and women held their property separately. Divorce was provided for on a number of grounds, for example impotence or homosexuality of the husband. Household property could not be disposed of without the consent of both husband and wife and the law restricted a husband's right to 'correct' (hit) his wife was limited with fines and a woman had the right to divorce a husband who left a mark on her (Ryan, 2010).

10. Film Censorship 1923, Refusal of Divorce 1925,Censorship of Publications Act 1929, Public Service 1932 Marriage Bar, 1935 Public Dance Halls Act, 1935 Control of Fertility in the 1935 Criminal Law Amendment Act, Conditions of Employment Act 1936, 1927 Juries Act.

11. A willingness and desire to dedicate oneself and venerate, in terms of prayers or in terms of a set of pious acts to the Virgin Mary. These prayers or acts may be accompanied by specific requests for Mary's intercession to God.

12. Mary Immaculate Teacher Training College, Limerick, Manual, 1927.

13. A term applied to a 'new breed' of young Western women who wore short skirts, bobbed their hair, listened to jazz and flaunted their disdain for what was then considered acceptable behaviour.

14. Magdalene asylums were institutions from the 18th to the mid-20th centuries ostensibly for 'fallen women', a term used to imply sexual promiscuity. The first asylum in Ireland opened on Leeson Street in Dublin in 1765, founded by the Protestant Lady Arabella Denny. Initially the mission of the asylums was often to rehabilitate women back into society, but by the early 20th century the homes had become increasingly punitive and prison-like. In most of these asylums, the inmates were required to undertake hard physical labour, including laundry and needle work punctuated by a daily regime that included long periods of prayer and enforced silence. In Ireland, such asylums were known as Magdalene laundries. It has been estimated that up to 30,000 women passed through such laundries in Ireland. The last Magdalene asylum, in Waterford, Ireland, closed on 25 September 1996.

15. In reality sexual assault/rape cases were rarely reported in the Irish press but when they were victims were given no anonymity (Keating, 2012a).

16. Thomas Hobbes of Malmesbury (5 April 1588–4 December 1679) was an English philosopher, best known today for his work on political philosophy. In *Leviathan*, Hobbes set out his doctrine of the foundation of states and legitimate governments – originating social contract theory much of the book is occupied with demonstrating the necessity of a strong central authority to avoid the evil of discord and civil war. Beginning from a mechanistic understanding of human beings and the passions, Hobbes

postulated what life would be like without government, a condition which he calls the state of nature. In that state, each person would have a right, or licence, to everything in the world. This, Hobbes argued, would lead to a 'war of all against all'.

17. Arensberg and Kimball's *Family and Community in Ireland* was a social anthropological study arising from an interdisciplinary research programme undertaken in Ireland in the 1930s known as the 'Harvard Irish Survey' (1931–1936). The major focus of the work was on County Clare.

18. The theme of the nobility of Irish manhood being worn away by years of privation was sighted by a Connaught jury as a basis for mercy for a man who had broken into homes to rape and beat two women on the same night (Connaught Telegraph, 14 February 1925).

19. O'Duffy took over command of the Army Comrades Association (ACA) in 1933 and renamed it the National Guard an organisation better known as The Blueshirt (Irish: *Na Léinte Gorma*), a right-wing Irish political organisation active in the 1930, more opposed to the Irish Republican Army and Sinn Féin rather than an overtly Fascist organisation. O'Duffy then founded the National Corporate Party, and later raised an 'Irish Brigade' that took Franco's side in the Spanish Civil War. A long-term admirer of Mussolini's Fascist Italy, O'Duffy became increasing drawn to Nazism towards the end of his life. He died in 1944.

20. He was later to describe himself as the 'third most important man in Europe' after Hitler and Mussolini.

21. A shebeen was an illicit bar or club where alcohol was sold without a licence.

REFERENCES

Arensburg, C. M., & Kimball, S. T. (2001). *Family and community in Ireland* (3rd ed.). Ennis: Clasp Press.

Beaumont, C. (1997). Women, citizenship and Catholicism in the Irish Free State. *Women's History Review*, *6*(4), 563–585.

Bourke, J. (2008). *Rape. A history 1860 to the present*. London: Virago.

Brown, T. (2010). *Ireland: A social and cultural history, 1922 to 2002*. New York, NY: Harper Perennial.

Cross, E. (1981). *Tailor and Ansty*. Dublin: Mercier Press.

Ferriter, D. (2005). *The transformation of Ireland, 1900–2000*. London: Profile Books.

Finnane, M. (2001). The Carrigan Committee of 1930–31 and the moral condition of the Saorstat. *Irish Historical Studies*, *XXXII*(128), 519–536.

Finnegan, F. (2004). *Do Penance or Perish: Magdalen Asylums in Ireland*. New York, NY: Oxford University Press.

Horgan, J. (1995). Saving us from ourselves: Contraception, censorship and the 'evil literature' controversy of 1926. *Irish Communications Review*, *5*(1), 61–67.

Howell, P. (2003). Venereal disease and the politics of prostitution in the Irish Free State. *Irish Historical Studies*, *33*(131), 320–341.

Inglis, T. (1998). *Moral monopoly. The rise and fall of the Catholic Church in modern Ireland*. Dublin: UCD Press.

Keating, A. (2002). *Secrets and lies: An exploration of the role of identity, culture and communication in the policy process relating to the provision of protection and care for vulnerable children in the Irish Free State and Republic 1923–1974*. PhD: Dublin City University.

Keating, A. (2012a). Sexual crime in the Irish Free State 1922–1933: Its nature, extent and reporting. *Irish Studies Review*, *20*(2), 137–158.

Keating, A. (2012b). The case against the Waterford Standard. *New Hibernia Review*, *16*, 2.

Kennedy, F. (2000). The suppression of the Carrigan Report: A historical perspective on child abuse. *Studies*, *89*(356), 353–363.
Larmour, L. (1998). *Aspects of the state and female sexuality in the Irish free state 1922–1949*. Unpublished PhD. University College Cork.
Luddy, M. (2007). *Prostitution and Irish society*. Cambridge: Cambridge University Press.
McDONALD, D. (2010). Attitudes to rape victims wrong, says McAleese. *Irish Independent*, January 18. Retrieved from http://www.independent.ie/irish-news/attitudes-to-rape-victims-wrong-says-mcaleese-26623934.html. Accessed on May 14, 2013.
McGarry, F. (2005). *EoinO'Duffy: A self-made hero*. Oxford: OUP.
McNiff, L. (1997). *A history of the Garda Siochana*. Dublin: Wolfhound Press.
Murphy, K. (2005, November–December). An Irish General: William Richard English Murphy, 1890–1975. *History Ireland*, *13*(6), 10–11.
O'Conner, B. (2003). Ruin and romance: Heterosexual discourse on Irish popular dance 1920–1960. *Irish Journal of Sociology*, *12*, 2.
Ryan, L. (2000). Drunken Tans, representations of sex and violence in the Anglo-Irish War. *Feminist Review*, Number 66.
Ryan, L. (2002). *Gender, identity and the Irish Press, 1922–37: Embodying the nation*. New York, NY: MellenPress.
Ryan, M. (2010). A feminism of their own?: Irish women's history and contemporary Irish women's writing. *Estudios Irlandeses*, *5*, 92–101.
Szmigiero, K. (2006). Fabulism and Irish censorship. *Estudios Irlandeses*, *1*, 112–118.
Valiulis, M. G. (1995a). Power, gender and identity in the Irish Free State. *Journal of Women's History*, *6–7*(4–1), 117–136.
Valiulis, M. G. (1995b). Neither feminist nor flapper: The ecclesiastical construction of the ideal Irish Woman. In D. O'Dowd & S. Wichert (Eds.), *Chattel, servant or citizen: Women's status in Church, state and society* (pp. 168–178). Belfast: The Institute of Irish Studies.
White, J. H. (1971). *Church and state in Modern Ireland 1923–1970*. Dublin: Gill & Macmillan.

Archival Sources

Garda Archives, Dublin.
Department of Justice Files. The National Archives, Dublin.

Newspapers

Conaught Telegraph 14.2.1925.

A SPATIAL ANALYSIS OF CRIME: 'THE WIRE' AND DEPICTIONS OF URBAN CRIME

Shane Leonard

ABSTRACT

This chapter sets out to examine the topic of a spatial analysis of urban crime through an analysis of David Simon's seminal television series The Wire. *By developing an analysis of the issues that are presented in the series, issues such as race, ethnicity and representation will be addressed in order to add to the understanding of these topics in relation to race and media representations. Each section will address a set of themes which are evident in* The Wire. *The chapter highlights the idea of race in the series and how characters are presented on screen. The research is also concerned with economic issues depicted in the series and the effect of the economy on the characters in Baltimore, the U.S. city in which* The Wire *was set.*

The conclusion of the chapter addresses poverty class and inequality as topics and sets out to document these themes in relation to race. The third chapter also discusses the racism and discrimination that is apparent in The Wire. *By contextualising the series, the book is attempting to theorise relevant issues surrounding race, gender and power through an examination of relevant literature and the development of a theoretical framework from which key issues will be addressed.*

Keywords: Spatial analysis; crime; media; 'The Wire'; depictions of urban crime; Baltimore

Environmental Criminology: Spatial Analysis and Regional Issues
Advances in Sustainability and Environmental Justice, Volume 20, 103–133

ISSN: 2051-5030/doi:10.1108/S2051-503020170000020007

INTRODUCTION AND THEORETICAL FRAMEWORK

This chapter examines issues of race and representation as depicted in the Home Box Office (HBO) television series *The Wire*. Set in Baltimore, Maryland, in the United States, *The Wire* sets out to examine how structures, hierarchies and bureaucracies operate in a contemporary urban setting. The racially diverse nature of Baltimore is successfully captured in *The Wire* which creates a number of fertile possibilities for academic discussion. The study will make use of academic texts, articles and a good knowledge of the series to provide a researched analysis of *The Wire*. The three chapters of the book will examine different fields of study and present a number of ideas discussing the content of the series. The chapter will focus on the theme of race and will outline how different ethnicities are represented in the series.

The chapter will attempt to engage with the economic aspects of *The Wire* and highlight how capitalism, globalisation and modernisation have a profound effect on the characters in the series. It will also discuss race but will be focused on how race affects issues such as class, poverty and inequality. While *The Wire* is obviously a more recent work, it is so sophisticated and realistic that it manages to provide a comprehensive understanding of Baltimore society which reveals a long-standing history of inequality and presents an anxious image of future issues which may unravel.

Kelly and Gay are utilised in order to help position the series in relation to the media and TV. Their articles on the meaning of representation of race help illustrate both the importance and the impact of different ethnicities in the series. Fraley's ideas on identity also document the significance of the depictions of the characters in *The Wire* and also show how these portrayals can be understood with regard to wider developments. Sharma positions the series not only in relation to wider TV, but also in relation to how it resonates in the academic community. Franklin and McNeil describe how the development of characters on screen is tied up with wider issues presented in the series.

Toscano and Kinkle illustrate the economic difficulties presented in the series and the attempt of *The Wire* to map out issues. Fuggle's article debates the idea that the series is an attempt to critique institutional power. Blauner's ideas on internal colonialism are used to help understand how urban developments have changed and what *The Wire* represents today. Meaney describes how characters struggle because of socio-economic issues in Baltimore in the series.

Weir and Bilton provide a theoretical grounding for terms like 'the underclass' and thus help highlight the connectivity of themes of class and race. Goldberg's understanding of race and racism assisted in documenting the type of discrimination which occurs in the series. Spencer's conception of ethnic difference in society also brought a level of depth and understanding to how racial issues in the series were presented. Castles highlights the changing nature of racism, the influence of class and how globalisation has influenced the role of race in society.

The main theoretical framework surrounding this study incorporates a number of key concepts from contemporary theory. Media theory and methods of representation ground the series to contemporary developments in this area. Political economy and the Left/Right ideological divide are utilised to establish a sense of contrast in the media and illustrates the role of the TV series. Concepts of globalisation, capitalism and neo-liberalism bring an academic depth to the book's arguments. Internal colonialism helps address ideas like gentrification, urban segregation and ghetto revolt. Gender analysis is also used to help establish the different roles of each character in the series. Ideas of racism and institutional racism document the different forms of discrimination. Theories of class also underpin developments in the series and add to the main argument in the text.

A SPATIAL ANALYSIS OF CRIME IN AN URBAN CONTEXT: *THE WIRE*

The main focus of this chapter will be an analysis of the idea of race and discuss media representations in an effort to develop understanding of the role of *The Wire*. By addressing this area the chapter sets up the idea of using a TV series as an academic study and establishes *The Wire* as a credible series which is worthy of analysis. Making use of articles including media portrayals will establish how *The Wire* often differentiates from stereotypical depictions. Lisa Kelly describes the series as an attempt to: 'create a complex, authentic, and credible depiction of the American city of Baltimore, through which to examine the themes of class, institutional corruption and corrosion, and the ongoing drug war' (Kelly, 2011).

Interestingly, the series depicts a number of different ethnicities living in the city of Baltimore. There are African-Americans, Irish Americans and Polish Americans along with a number of sub cultures. What separates this series from others is the ability to provide realistic and authentic depictions of characters. Issues within the city of Baltimore are also captured in a very strategic way which introduces the viewer to a unique setting. The series has gained popularity within the academic community for presenting issues and characters in a balanced or considered manner. *The Wire* is one of the most notable series of the 21st century with regard to its ability to influence opinion. Those living in the projects, the homeless or people who have been caught up in crime are often portrayed in a manner which provokes empathy. As *The Wire* provides the point of view of the police and their targets, notions of good and evil are often challenged.

SPATIAL REPRESENTATION AND *THE WIRE*

The Wire earns its merit by ensuring its portrayals are accurate, the casting in the series is carefully considered so that an authentic depiction is guaranteed. Kelly highlights how accurate casting is a positive feature of the series: 'What makes *The Wire* even more distinctive, however, is the fact that it features a cast that is around sixty-five per cent black' (Kelly, 2011). While the series attempts to pay attention to a number of different communities in Baltimore, it could be argued that its primary representative focus is the African Americans in the city. However, what must also be considered is the target audience when discussing the series. *The Wire* certainly stands out amongst other HBO (Home Box Office) series which would generally rely on a predominantly white cast. One may also be inclined to think that the largest demographic of people watching HBO series, including *The Wire*, would be predominantly white. This creates some issues with regard to the impact of the series as a piece of representation. Even if the director has an ability to create an accurate portrayal, the content should be available to a wide range of people, especially communities that the series may depict.

However, even considering this issue the series still retains validity because of its original nature. The consistent casting in *The Wire* gives the series a convincing appeal and also creates an interest among the academic community: 'the casting process in *The Wire* affirms the sense of "otherness" that informs the show; the "other" American that tends to go unseen on television screens due to the focus on mainly white, middleclass and ultimately "aspirational" characters' (Kelly, 2011). It could be argued that *The Wire* is in fact one of the first TV series that attempted to portray African Americans in a positive, or at least considered light. While this is not to say that other depictions were necessarily racist in nature, one could claim that media representations often neglected African Americans in the past. Speaking of the casting choices of the series, Kelly states that *The Wire* features characters that are ultimately playing: 'versions of themselves in an attempt to lend authenticity to the conceit of each show' (Kelly, 2011).

There are a few important things to remember with regard to Kelly's insight here. Firstly, one should be careful to remember that while the series obviously portrays some of the African Americans in low social positions, the series is supposed to be an accurate portrayal of what is a predominantly poor city. *The Wire* does not solely depict African Americans as impoverished members of the projects. In the series African Americans also take up roles as policemen, lawyers, political representatives and religious figures. However, while some characters in *The Wire* are able to climb the socio-economic ladder of Baltimore, it could be claimed that some of these characters are exceptions to the restrictive nature of the city.

The Wire still seems to be one of the few exceptions when considering the larger domain of television. 'With regards to the prime-time television landscape, white cast members continue to "dominate" "screen time," not only in terms of the number of characters, but also in terms of the *prominence* of the characters in their respective shows' narratives' (Kelly, 2011). The wider misrepresentation in TV series is potentially damaging as it creates strain among different ethnicities and imposes boundaries among people.

In an article titled 'Prime Time Apartheid: how television has divided the races', Jason Gay argues that the seemingly stratified world of television is not so much racially influenced, but commercially dictated. Gay claims that: 'Racism alone doesn't explain the current state of African-Americans on TV; more than anything else, the marketplace does'. It seems that the direction of TV is changeable and that audiences want different content at different times. One would be inclined to believe that wider political developments may affect viewing patterns or openness to unorthodox depictions. 'Whereas previous decades saw programmes such as *The Cosby Show, The Jeffersons* and *The Flip Wilson Show* attract broad and racially diverse audiences, the 1990s have been a decade of increasing racial division in the ratings'. It is interesting to note how viewing trends and methods of representation can change drastically from one decade to the next.

While previous TV series depictions featuring African Americans may have been attempts to cash in on fluctuating trends among viewership, *The Wire* presents itself as above purely chasing commercial success. The series is primarily concerned with quality portrayals and accurate representation of the unheard. One could argue that *The Wire* was the first drama depiction that featured African Americans that achieved popular, critical and commercial success. The usage of Baltimore locals may have bolstered the series grass roots charm: 'The often impenetrable accents and highly specific colloquial vernacular make no concessions to the average viewer but instead demand their attention and commitment' (Kelly, 2011).

While viewers may not initially be able to understand characters because of localised slang or accents, the viewer is encouraged to trust the series fast paced dialogue. *The Wire* promotes critical thinking and encourages viewers to engage with the depictions in a constructive manner. The series succeeds because it does not fall into familiar pitfalls of less accurate depictions, *The Wire* never relies on stereotypes, avoids lazy writing and overcomes viewership limitations. Praising director Simon's ability to 'cast with a conscious', Kelly states that: '*The Wire* would be a very different programme if the supporting cast were filled with jobbing actors rather than Baltimore residents with first-hand experience of the drug crisis within the city' (Kelly, 2011).

REPRESENTATIONS OF POLITICS AND IDENTITY

Todd Fraley emphasises the role of representation on individual conceptions of identity: 'Popular media representations, serving as sites of contestation regarding the social construction of race, play a key role in shaping, communicating and understanding racial identities' (Fraley, 2011). In his article, titled 'A man's got to have a code: Identity, Racial Codes and HBO's The Wire', much of our own perceptions of race may be largely influenced by what we see in the media. Fraley also praises the series for its ability to address areas such as identity. Interestingly, *The Wire* seems capable of reconstructing the capabilities of TV as a credible medium. Through TV series like *The Wire*, people begin to reconstruct their notions of identity, race or representation.

Fraley continues by stating that while the majority of TV may focus on personal issues or individualistic desires, *The Wire* concentrates on wider sociological and political developments (Fraley, 2011). Perhaps the political nature of the show, alongside its social sensitivity, is what often qualifies *The Wire* for academic analysis over other series. In fact, it seems that the series itself stands out among the medium of TV: 'So while for *The Wire*, television and contemporary public media generally, are part of the problem of neoliberal culture, the TV series itself has been particularly valorized as an exception to the everyday banalities of television culture' (Sharma, 2011).

Detailing *The Wire* and its consistent ability to depict characters and their struggles, Fraley states that the series: 'presented complex individuals struggling with the inherent tensions of good and bad and rejecting simple racial categorizations' (Fraley, 2011). Even considering that the show heavily features policemen, lawyers and other political powers the series is far removed from more traditional American cop shows which promote themes of justice and order. One could argue that the series is more concerned with allowing people to sympathise with characters that have been marginalised. While a conventional depiction of policemen in an urban setting would usually rely on a villainous, disenfranchised character, *The Wire* equalises the notions of good and evil and represents who would otherwise be criminals as products of a difficult environment. At times characters are often presented as being potentially successful, but just lacking integration and acceptance from wider society in Baltimore. Comparing *The Wire* to the series *Hill Street Blues*, Fraley claims that *The Wire*: 'offers a similar analysis through tragic heroes and principled deviants who illustrate the tension between structure and agency' (Fraley, 2011).

It is consistently implied by those who have written on *The Wire* that many of the issues present in the series somehow relate back to racial tensions. 'For five seasons *The Wire* used gripping portrayals of complex social issues to interrogate the daily interactions of disconnected communities in a society wrestling with racial identity' (Fraley, 2011). This is an interesting insight into the current state of racial relations in America today. Disparities between rich and poor are further exasperated alongside issues of race. Evidently, this difficult situation results in problematic social issues which surface in the form of drugs, racism and poverty.

The series original depictions and authentic casting could be highlighted as the reason why the viewer must always reconsider and revaluate when watching *The Wire*: 'Because nearly every character played with stereotypes, each episode of *The Wire* demanded reconsideration of assumptions about authentic racial identities' (Fraley, 2011). Interestingly, in the series characters are equally subject to social issues regardless of their class standing. From the drug peddlers to the aspiring political figures, no character in *The Wire* is completely sheltered from the underlying issues in the city. One of the most impressive aspects of *The Wire* is its ability to blur boundaries between class and race and present characters on a humane and relatable manner. Interestingly, while the series shows the interconnectedness of society, it also manages to highlight social divisions without contradicting itself.

A political analysis must be deployed when considering *The Wire* as a piece of representation. It would be uncontroversial to claim that the wider political left and right divide trickles into media such as TV, movies and news channels. When one examines the content and messages of a TV series they may unravel the underlying political tones, and the resulting political affiliation. *The Wire* sets out to differentiate itself from previous attempts of representation. Traditionally, TV series would have generally depicted African Americans in an unsympathetic manner. Black and white characters were presented as occupying two separate worlds. This kind of representation would be considered right wing. What is meant by this is that the message in this kind of representation is consistent with the status quo of right wing parties and ideologies. The right wing message could be understood as promoting responsibility for ones actions, self-reliance and hard work. However, these messages are interesting to consider with regard to race and representation in popular media. At times the series presents a political message by using its ability to critique various institutions: 'Further, the series has an explicit political agenda and offers a set of critiques of neoliberal institutions and capitalism that resonates with much analysis of left-liberal academic circles' (Sharma, 2011).

The idea of having responsibility for one's actions may seem fine in an isolated sense but this discredits wider forces at play. Similarly, self-reliance shifts the responsibility to the individual and hence frees society or the state of any moral obligations. The promotion of hard work also fits in with the idea that nobody else is responsible for your actions. In contrast left wing ideology is concerned with context, equality and opportunity. By examining a number of different representations through TV, movies or literature one can develop a better understanding of the left/right divide in media and politics. When one studies *The Wire*, the left ideology becomes increasingly evident. In the context of America, the issue of right wing representation is that it denies history. What is meant by this is that right wing ideology may attempt to create marginalised citizens into ideal patrons, but there is a fundamental denial of an unfair political structure, a colonial past and a racial dilemma. TV representations can have a negative impact on the image of a particular ethnicity. Media may set out to disproportionately represent particular classes and portray people in a certain manner to influence public opinion. The underclass of a particular society may be over represented in media to spread a sense of moral panic.

In this scenario some of the black population may be being depicted and they may receive substantial air time, however, the undertone is negative and there is an underlying agenda to characterise these people in a negative light. By establishing this kind of image of African Americans, media could be accused of fuelling white supremacy. Similarly, a positive and more acknowledging view of African Americans may be denied and this would separate the two different ethnicities in terms of culture and understanding. This act of discrediting African American communities also reinforces the idea that there is only one homogenised and dominant culture. In America this culture could be identified and associated with being white, middle class and probably conservative. Comparing *The Wire* to other series such as *The Cosby Show*, and their ability to represent, Fraley claims that *The Wire* utilises: 'issues and offers a prime example of how TV can change a hegemonic social and racial order built upon separation by moving beyond constructed dichotomies and revealing the intersection and interdependence of class, race and gender' (Fraley, 2011). Again we are reminded of the potentially unifying abilities of a TV series such as *The Wire*. The series makes use of issues in a way that brings people together under the common banner of humanity as opposed to other series that may aim to create social segregation by vilifying different communities or ethnicities.

Considering that the media, particularly TV in the 1990s, was hampered with negative stereotypes, racial intolerance and cultural separation, *The Wire* has genuinely created a new landscape for more positive depictions and balanced representations. A genuine interest in providing quality content has given the series an undeniable appeal: '*The Wire* has overcome the limitation of questions concerning authentic, positive and negative, with a commitment to social realism' (Fraley, 2011). The series has the unique ability to change people's minds and can even redefine the capabilities of TV: '*The Wire* offers a blend of the contested meanings of race, and recognizes that racial experience is not singular' (Fraley, 2011). Fraley highlights how the series can capture unique qualities in characters and relate them to wider social trends. Stringer Bell, as an African American character, is an interesting example of how the series attempts to portray characters in an authentic manner.

CULTURAL DIFFERENCE AND THE ETHNIC COMMUNITY

Even considering Bell's involvement with the drug empire in Baltimore, his character develops aspirations to become legitimate in both his personal and commercial life. However, what is interesting about this case is the fact that the character pursues what would be considered the American ideal of life. Bell hopes to become more business-like, approachable but most importantly accepted, particularly by the white community. This example provides an interesting insight into the complex identities of the characters in *The Wire* and it seems to exemplify some of the cultural transitions in ethnically diverse cities such as Baltimore. Franklin claims that the series features many dynamic characters that often have to adapt their morals to survive: '*The Wire* offers a great many such transitional figures: people whose principles are at odds with the organizational systems in which they find themselves' (Franklin, 2011). Franklin also shows an awareness of how the series presents race as a key factor effecting how the characters survive and interact in Baltimore: 'The people within each institution in *The Wire* are shown perpetually subsuming racial difference beneath the effort to survive against the system' (Franklin, 2011).

However, with this in mind one may feel the need to criticise the homogenised nature of this character development. Stringer may be pursuing a different lifestyle in the series but he aspires to essentially white middle class life. This aspect of the series seems to represent post-colonial critique in a contemporary way. The series depicts Baltimore as only appreciating one culture, one which is predominantly commercial, competitive and homogenised. Stringer is essentially subjects to American domination; the system, the authorities and the class system all pressure Stringer into adopting more normalised and accepted behaviours. This is not to say that Stringer's drug involvement, amoral actions and crimes should be applauded; however, one could argue that the character is only left with one alternative, one correct method of socialisation. Interestingly, *The Wire* does not fall into the trap of portraying only African Americans in this moral dilemma. Season two heavily features the Polish American community who struggle to find themselves between legitimate and illegitimate spheres.

Omar Little who is one of the most interesting characters of the series anchors the shows attempt to redefine racial identities. '*The Wire's* challenge to extant racialized codes is clearly articulated in Omar's belief that: "A Man's gotta have a Code" which provides an entry point to understand how the characters in *The Wire* contest historical representations of black and white and a media history requiring individuals to be situated in a black/white dichotomy' (Fraley, 2011). Omar's motto on the importance of individual moral codes helps the viewer understand the sociological landscape of Baltimore in the series, while providing a window into the moral subtext of the series. Fraley states that: 'Popular media representations, as symptomatic texts require further analysis regarding their response to a social reality containing fluid and unstable identities posing challenges to traditional boundaries creating social cleavages' (Fraley, 2011). In this instance, it seems the ability of *The Wire* is to portray characters of different ethnicities struggling with similar issues whether they may be personal choices, ethical dilemmas or ideological outlooks.

The Wire could be identified as a type of representation which can lead to progressive thinking: 'Over the course of five seasons, *The Wire* highlighted the complexities of identity and race, and placed "other" America at centre stage to compel privileged/default America to examine the consequences of misguided and shortsighted codes' (Fraley, 2011). The series is concerned with race, but specifically with consequences of ethical difference in a diverse city like Baltimore. *The Wire* also succeeds in repositioning TV as a potentially thought-provoking and intellectual medium. One could argue that the series presents blackness as the norm in the city of Baltimore. The series allows the viewer to compare and contrast Blackness and Whiteness in a critical manner: 'The unusual, numerically significant presence of an African-American characters through the series, across institutional and class lines, arguably makes blackness the norm' (Sharma, 2011). If the viewer can establish difference or similarity between White and Black developments in the series it would also seem relevant to create an understanding of how issues are impacting different ethnicities.

The depiction of the Polish American community in season two of *The Wire* is crucial to the series claim of authentic depictions. Daniel McNeil's 'White Negroes and The Wire' addresses the mixing of black and white culture in a contemporary context. McNeil highlights the cultural capabilities of the series: 'We do not have to limit ourselves to the deconstruction of a creative class or a post-racial meritocracy when *The Wire's* second season plays such an eloquent eulogy to the White Negro and "the Wigger" in a working-class Polish American community' (McNeil, 2011). *The Wire* does not limit itself to only addressing African American representation. After all, the series sets out to explore issues affecting the entire city, not just one part of the population. The representation of the Polish American community not only adds depth to the series, but also creates a sense of racial equality with regard to depiction. The series didn't want to only portray African American characters; it also wanted to avoid only depicting African American struggles, as it is ultimately an American tale.

The utilisation of the Polish American community sets up for a certain amount of comparing and contrasting in the series. It becomes evident that both these communities are struggling to find themselves in Baltimore. Evidently, one would be inclined to wonder if these issues are more linked to class than race. Franklin claims that the theme of race is always present in the series, but it is not necessarily always the most important: 'Race is always as presence and an issue, but it is subsumed in the bigger problems of the social environment' (Franklin, 2011). For example, Frank Sobotka, who is a big figure in the docks industry in Baltimore, could be described as colour-blind in the sense that often sees such issues as being rooted in class difference and not racial difference. The unique manner in which *The Wire* attempts to represent also transforms definitions of race throughout the series: 'The normalization of blackness and "making whiteness strange" is progressive in terms of the politics of representation, but the theoretical effect of these deconstructions makes race either marginal or invisible to the politics of the show, if we remain at the textual level' (Sharma, 2011).

McNeil uses Mailer's *The White Negro* as a backdrop to his discussion of the Polish Americans in *The Wire*. He summarises Mailer's attempt to categorise the identification of the African American community in modern America. Mailer is described as setting out to explore the idea that: 'Blacks were ostracized from American mainstream, but choosing to identify with the plight of urban African Americans rather than ascribe guilt or shame' (McNeil, 2011). Season two of *The Wire* is focused on very real issues such as unemployment, income inequality and illegitimate trade. One must remember that while the Polish Americans may be a white community, they are still a fringe population with regard to the wider demographic. 'The second season of *The Wire* sought to capture the last days of being able to follow in your old man's footsteps to make a living if nothing better – and legitimate – came along' (McNeil, 2011). One could claim that the drug empires and the dock workers are plagued by the same issues. Both communities struggle to find legitimate employment and must break the law in order to have an income and a standard of living. The season, in the words of the creator, David Simon, was: 'a twelve episode wake for the death of work' (McNeil, 2011).

It becomes increasingly evident that *The Wire* is a critique of contemporary mainstream America. McNeil summarises the series intentions, particularly in season two: 'this eulogy for working-class Baltimore was inspired by White Negroes and Black Power activists, a link made explicit by Simon's attack on people and desiring machines he calls "soulless white guys," the champions of computers, condominiums and corporate takeovers' (McNeil, 2011). What is interesting about this depiction is that similar to the African American drug dealers portrayed in season one, there is supposed to be a certain amount of empathy associated with these characters. McNeil describes the dynamics on screen during season two and addresses one of the main figures in that season, Frank Sobotka: 'Sobotka is compromised by his dealings with businessmen involved in the trafficking of drugs, weapons and women, but rationalises his participation as a means to support individual workers who suffer financial hardship or injury and obtain a canal vital to their trade' (McNeil, 2011). The season seems to encapsulate working-class Americans who are struggling to compete in an increasingly globalised trading network. Both technology and multinationals are highlighted as potential issues for the dock workers who find their methods becoming dated. 'Without the canal, Sobotka is well aware that his livelihood will be threatened by property developers and a downsized port governed by technocrats' (McNeil, 2011).

Some of the Polish American characters in the season are left culturally confused and seem to have somewhat inconsistent identities. McNeil highlights Ziggy, who is Sobotka's son, as the estranged product of the barren docklands: 'Left with a tropical shirt and a colourful hat, Ziggy is doomed to live like a tourist in his place of birth, frustrated with the creased shirts of his father's friends and the over-sized tops of the Blacks in outlaw spaces' (McNeil, 2011).

A SPATIAL ANALYSIS OF GLOBAL CRIME IN LOCAL SETTINGS

Following on from Series One and the depiction of local youth, and the depiction of the Polish American community struggling to adapt in season two, it seems fitting that this analysis should deal with themes such as globalisation, capitalism and neo-liberalism and the affects on the characters in the series. This perspective is also interested in developing an understanding as to how structures in Baltimore operate, and how they may stifle progress in the city. Toscano and Kinkle's online journal 'Baltimore as World and Representation: Cognitive Mapping and Capitalism in The Wire' gives us a unique insight into the socio-economic developments in the series. The article mentions the series ability to cover a wide range of issues: 'While the first season largely revolves around the drug trade, subsequent seasons expand the scope of the show to cover de-industrialisation, city hall, the school system and the media' (Tosacno & Kinkle, 2009).

Like other writers on the series, Toscano and Kinkle go out of their way to praise the series interconnectedness and the networked way in which developments are mapped out. The article expresses interest in the way Baltimore is portrayed in the series, issues like unemployment are explored in a unique manner: 'the evaporation of working class jobs leads young men into the drugs trade' (Tosacno & Kinkle, 2009). The economy of Baltimore that is presented in *The Wire* is its own entity which has a profound impact on the characters in the series.

DISENFRANCHISEMENT AND THE POLITICAL ECONOMY OF CRIME

Toscano and Kinkle show an awareness of the series wider hopes and aims. Issues are not presented in an isolated sense, the series always attempts to provide context and to link the content together: 'The show descends into the hidden abode of street-level drug distribution, not to merely depict the violence and hopelessness that exists in these neighbourhoods, but to expose their complex organisation and their hostile yet symbolic relationship with the state and neoliberal institutions' (Tosacno & Kinkle, 2009). What is interesting about *The Wire* is that it seems to cover a wide number of issues, but also present them in a consistent and detailed manner. While the series focuses on a number of themes like race, politics and identity, there is still room for an economic aspect within the episodes: 'The themes of relentless devaluation, dispossession and decline are writ large' (Tosacno & Kinkle, 2009).

Simon has made large claims with regard to both the capabilities of the series and also the state of American politics and society. He has described the show itself as a portrayal of 'the decline of the American empire' and he is also adamant that the current political and social structures in place are not serving contemporary America in a progressive manner. In this sense one could claim that the show has a political agenda, however the themes of the show are constructed so intelligently, and the characters so realistic that the viewer has no choice but to trust the series. The series seems to have the ability of characterising structures and institutions. 'In *The Wire* the weight of the system, which we could perhaps curiously define as late capitalism and its institutions, assumes casual agency as an antagonist' (Tosacno & Kinkle, 2009). Similarly, it is made evident that the series is more relevant when examined in light of actual issues: 'we want to hold together, and if need be in tension, the "picture" of the urbanisation of capital will fully projected by *The Wire* with a broader reflection on the aesthetic and epistemic challenges of "mapping" or "representing" the contemporary capitalist world' (Tosacno & Kinkle, 2009).

Toscano and Kinkle highlight the importance of 'mapping out' real world issues, they claim that by using this process one can create greater understanding of what needs to be done and what needs to change. With regard to the series and cognitive mapping, they claim that: '*The Wire* can be understood as one of the most cogent attempts at producing a work classifiable under such an aesthetic' (Tosacno & Kinkle, 2009). In the article there is also a discussion of ideologies and efforts made to bring light to developments. One should consider a quote by Brian Holmes which is used in the article 'the major intellectual project of the worldwide Left in the 1990s was to map out the political economy of neoliberal capitalism, which had literally produced a new geography'. This quote seems to align with some of the content of *The Wire*. There is a consistent and obvious effort made to challenge institutional power and to redefine contemporary structures in the series.

What is interesting about Baltimore and its population, as presented in the series, is that while the mechanisms of globalisation and capitalism have such a striking influence on the dynamics of the city, the inhabitants themselves seem somewhat unaware of a world outside Baltimore. 'There are very few scenes which take place outside of Baltimore, and almost all these depict the Baltimorean out of their element' (Tosacno & Kinkle, 2009). This raises a number of interesting angles which could be explored regarding the depiction of this type of confined existence, urban geography and education. One should be careful to remember that like any other aspect of the series, it is still a portrayal, and an interpretation rather than real-life documentation. However, with this vision of reality is underpinned by other issues like drug use, unemployment and poverty which are presented throughout the series. Baltimore city seems to act as a type of urban prison as it is depicted in *The Wire*. The characters in Baltimore are portrayed as being restricted both geographically and culturally to the boundaries of the city. Similarly, one would be inclined to believe that opportunities to escape out of the impoverished areas of Baltimore, particularly the projects would be quite rare.

On a practical level it seems evident that the unskilled and uneducated characters depicted in *The Wire* would most likely find themselves struggling in other parts of America. Because they have no conventional skills or training to offer they would more than likely fall prey to similar illegal trade networks or drug dealing but just in another context. Interestingly, this cultural or societal deficiency is not only limited to the lower classes presented in the series. It seems that nearly everyone in Baltimore is subject to this contained type of living. As mentioned in the previous chapter it is essential to remember the series positioning among wider forces. Politically, the series is could be identified as a series that highlights common humanity and attempts to examine power structures as they exist. However, it is also important to remember its positioning with regard to economics. Toscano and Kinkle state that *The Wire* makes no attempt to cover up its agendas or aims. The fact that the show may at times affiliate itself with ideological structures gives it an academic interest and appeal.

Speaking of the economic influence in the show, one could claim that the series is: 'a kind of labourist social critique, infused by a dose of nostalgia for the Fordist compact' (Tosacno & Kinkle, 2009). It is interesting to note that *The Wire* attempts to shift the blame of unemployment away from the people of Baltimore and onto the institutes, the power structures and local bureaucracies. In the second season of *The Wire* it is interesting to see the effects of globalisation on local developments in Baltimore and other social exchanges which happen in the neoliberal setting. However, another aspect of the series is the globalisation of crime. In this way this is where season one and season two differentiate. Season one is focused on showing personal issues and struggles in a confined setting. However, the second season, in many ways, extends the scope of the damage of the illegal trade networks. The second season reveals a new aspect of Baltimore but also opens up a more globalised angle to the show. Simon describes the second season as: 'a multi-dimensional study of the grinding downsizing of the Baltimore docks and their articulation with global flows of capital' (Tosacno & Kinkle, 2009). While *The Wire* may not be directly blaming globalisation for the issues in the city of Baltimore, it is still interesting to see how an impoverished city's dysfunctions can have such effects in other areas.

This representation of people losing out on work because of increased use of technology is one of the most accurate depictions in the series. Those who have written on the economic trends in Baltimore city have expressed concern for the future of the city in the post-industrial age: 'David Harvey, Marxist geographer and long-time resident of Baltimore indicates, the city lost two-thirds of its manufacturing employment after 1960' (Tosacno & Kinkle, 2009). The industrial nature of the city is particularly problematic because there is a lack of indigenous economy in Baltimore. The drug trade is one of the few trading networks that is both operated and utilised by the people of Baltimore. Toscano and Kinkle claim that *The Wire*: 'has a lot in common with a nostalgic valorisation of the moral economy of work and craft' (Tosacno & Kinkle, 2009).

One could state that the series portrays workers as struggling to find fulfilment in their work. Even those who manage to find consistent employment are subject to the wider industrial demands of the city and cannot achieve social mobility. The series explores issues within modernity regarding production and industrialisation but the aim is to show how the system no longer operates to serve the people. The citizens in Baltimore could be viewed as input into the labour process, unable to gain a proper foothold on capital or a part of the economy.

Essentially, *The Wire* does not stop at just critiquing the economic systems in Baltimore. There is a wider analysis of power structures and the legal system which prompts the viewer to believe institutional hierarchies are not serving the citizens, they are in place to protect agendas. Sophie Fuggle's article 'Short Circuiting the Power Grid: The Wire as Critique of Institutional Power' suggests that *The Wire* significantly contrasts with other shows dealing with law and order. While *The Wire* may represent authorities as being confused and undecided with regard to the best approach to dealing with crime, shows like *CSI* and *Law and Order* glamorise: 'the criminal justice system as a slick and efficient machine which upholds absolute values such as "truth" and "justice" and in doing so fully subscribe to such discourses' (Fuggle, 2011).

David Simon is often credited for his ability to reveal connectedness between different parts of society. Interestingly, in *The Wire* the various authorities police, detectives and lawyers are shown as representing one wider institute of power: 'Individual departments such as homicide, narcotics and major crimes are not depicted in isolation from one another but rather combine along with local police precincts, forensics, the district attorney's office and the courthouse to form a grid or matrix of disciplinary power' (Fuggle, 2011). The series has the ability to highlight some of the darker aspects of society regarding survivalist nature and individualism. Many characters presented in the series are left fending for themselves, even regardless of their standing in society.

It is interesting to consider this seemingly unified and cohesive body of various legal representatives. One would imagine that such connectedness in this context would seriously curtail the amount of crime in the series. However, it seems that the opposite is in fact the case, Fuggle discusses the confusion which emerges from this power structure: 'a tangled web of bureaucracy and confusion with personal rivalries and hierarchies, budgetary issues and questions of protocol preventing different departments from working together effectively' (Fuggle, 2011). While other TV series may vilify minority groups, the underclass or criminals, *The Wire* points out flaws among people in positions of power. Again, in this instance *The Wire* proves to be an accurate depiction, the police forces effort to manipulate stats in order to please bureaucratic demands proves to be an actual problem in Baltimore: 'junking the stats in *The Wire* has been a genuine problem in real-life Baltimore where the police department came under fire in 2006 for manipulating crime rate statistics' (Fuggle, 2011).

One point of interest which is to be found in *The Wire* can be seen in the journey of development undertaken by the series' characters. Many of the characters in the series have a very personal experience with the institutions of power which engulf them, based on issues such as race or gender. Again, this would point to the authenticity and credibility of the series. If one would imagine a modern city in America, it would be likely that many people would have a variety of relationships with the wider forces at play, while still having to deal with the restrictions of race or gender. Even those born into similar circumstances will make different personal choices which will effect their interaction in society. In the series the more tactical characters are able to identify power structures in Baltimore: 'Regardless of their skin colour there are certain characters who buy into what, for want of a better term, we might call "white" discourses of sovereign power – they believe they can "possess" power and change thing for the better using this power; Tommy Carcetti, Jimmy McNulty and Russell "Stringer" Bell would be key examples' (Fuggle, 2011). With regard to Fuggle's quote here, one would believe that there are certainly characters who aspire to develop themselves and progress in the society. However, there are times when race is irrelevant with regard to social mobility; characters may simply be aiming to climb the socio-economic ladder.

INTERNAL COLONIALISM AND URBAN SEGREGATION

In the mid-20th century colonial theorists came up with the term internal colonialism to describe uneven development between different regions in one state. One could claim that *The Wire* depicts a form of this internal colonialism throughout the series. With regard to internal colonialism, the exploitation generally involves manipulating the minority groups in order to serve capitalist demands. Also, while this may be the case in the TV series, it is interesting to consider the theory applied to the contemporary American state. In an article titled 'Internal Colonialism and Ghetto Revolt', Robert Blauner's book explores and develops the idea that: 'white-Black relations in America are essentially those of colonizer and colonized' (Blauner, 1969, p. 393). While the date of publication is obviously quite dated, it is interesting to see how theorists attempted to organise these developments in America. Blauner attempts to differentiate: 'colonization as a process and colonialism as a social, economic, and political system' (Blauner, 1969, p. 393). This insight provides some understanding with regard to the definition of contemporary American society. While America may not be in a phase of actual colonization, there are still colonial legacies haunting the country.

Blauner continues by stating that while the treatment of African Americans may in America may be similar to the treatment of non-whites around the world, the setting in which the discrimination occurs is quite different from typical colonial exchanges (Blauner, 1969, p. 393). Even considering the early date of this publication, Blauner highlights the need to re-examine the positioning of African Americans in American society. Contrasting a number of different ideas and perceptions regarding the Black population in America, Blauner claims that: 'Black people as an underclass are to a degree specifically exploited and to a degree economically dispensable in an automating society' (Blauner, 1969, p. 393). With this in mind, it is interesting to consider the African American population and the depiction of Black drug dealers in *The Wire*. Season one heavily portrays the operating drug empire in Baltimore which is essentially operated by African Americans. While the small time dealers are subject to the power of the more influential members of the network, it is evident that there is a power hierarchy at play. One could claim that there is a class system within this class system, from the highest ranks (Stringer Bell and Avon Barksdale) to the middle men (D'Angelo Barksdale) to the bottom rank dealers who simply operate in the physical movement of drugs around the housing projects.

However, with this in mind this system seems to originate out of a lack of opportunities and resources. The projects act as boundaries to the outside world and residents are left struggling for power, respect and a standard of life in a barren post-industrial landscape. Considering Blauner's discussion of the underclass being dispensable to the society, it seems that *The Wire* depicts this class as being almost banished to the projects where they are constrained and removed from the rest of the society. Even considering the date of the article Blauner highlights how traditional theoretical tools have not led to progressive action in America with regard to the development of African Americans. Similarly, Blauner claims that the only way to allow for social mobility and racial equality is to construct new frameworks with which to understand society. While *The Wire* captures the uncompromising nature of the projects it also manages to highlight the strategic nature of the drug empires and their operatives. The actual dealing of drugs is not praised by the series but the viewer cannot ignore the efforts to succeed in a difficult environment.

Blauner continues by documenting the colonial comparisons which have fuelled racial movements in the mid- to late-20th century: 'During the late 1950's, identification with African nations and other colonial or formerly colonized people grew in importance among Black militants' (Blauner, 1969, p. 394). This culture of resistance which is embedded in colonial history echoes into more recent times in the form of protest and is at times captured in *The Wire* when characters speak out against oppression and inequality. It is from these historical developments that the notion of internal colonialism originated: 'the U.S. was increasingly seen as a colonial power and the concept of domestic colonialism was introduced into the political analysis and rhetoric of militant nationalists' (Blauner, 1969, p. 394).

What is interesting is that one could claim that there is a link between the same mechanisms which were evident during phases of actual colonisation, the treatment of African Americans and the poverty and exploitation highlighted in *The Wire*. There is a definitive link between European colonialism and slavery in America: 'The essential condition for both American slavery and European colonialism was the power domination and the technological superiority of the Western world in its relation to peoples of non-Western and non-white origins' (Blauner, 1969, p. 396). One could stretch this analogy to the hierarchies which become evident in *The Wire*. The police force and other legal authorities hold a technological advantage over the criminals, drug peddlers and projects residents. In some ways one could view the police as oppressors and the citizens as oppressed in Baltimore, as depicted in *The Wire*.

This raises significant questions with regard to resistance and insurgence from the marginalised and disenfranchised characters in the series. It is interesting to see how infrastructural or technological advantage can foster racist thinking: 'This objective supremacy in technology and military power buttressed the West's sense of cultural superiority, laying the basis for racist ideologies that were elaborated to justify control and exploitation of non-white people' (Blauner, 1969, p. 396). While there may not be direct discrimination which is racially motivated in the series, there are many instances where characters struggle to break free from exploitative relationships.

Continuing the link between European colonisation of foreign territories and American slavery, Blauner states that: 'a common process of social oppression characterized the racial patterns in the two contexts despite the variation in political and social structure' (Blauner, 1969, p. 396). So it would seem that Blauner is making the point that colonialism, or at least its effects, is not strictly speaking an act which must involve two countries. He is essentially stating that internal colonialism can occur and that the effects of colonialism between a mother country and a colony can be seen within one domestic area. Relations of oppressor and oppressed, coloniser and colonised or powerful and powerless may emerge under the setting of internal colonialism and it seems that these relations also surface in *The Wire*.

GENDER AND POST-INDUSTRIALISATION

Gerardine Meaney's chapter, 'Not Irish Enough? Masculinity and Ethnicity in *The Wire* and *Rescue Me*' examines themes of class and race alongside one another. One of the focuses of her article is Jimmy McNulty, the most prominent Irish American character in *The Wire*. She identifies that: 'McNulty's Irishness is part of the season's exploration of white, working-class, masculine identities in which ethnicity is linked to the experience of immigration and assimilation' (Meaney, 2007, p. 10). It is important to examine the Irish American characters in Baltimore to add understanding to the different ethnicities in the show. Meaney claims that America is a country with distorted borders, she describes season two: 'The story's focus on the docks makes it a story of how America was and is made at the point of intersection between it and its others' (Meaney, 2007, p. 10). Commenting on the state of national security and economic instability in America is the early 2000s, Meaney states that pre-industrial sectors are struggling in a post-industrial society: 'The Polish dockworkers are becoming obsolete, a highly organized masculine labour force competing for work that is in short supply' (Meaney, 2007, p. 10).

Season two of *The Wire* perfectly captures the level of despair in an economically peripheral city like Baltimore. It seems that the main focus of this season is to portray a number of different ethnicities as struggling with global forces and the changing ways of the city. In this sense it would seem that the economy is to blame, so no individuals are vilified, at least in season two, rather everyone is portrayed as subject and at the mercy of the post-industrial economy in America. Speaking of how characters are affected by wider forces in the season, Meaney states that: 'In its dissection of the way in which global capital destroys traditional masculine and social roles, *The Wire* engages in an exploration of the causes of crime highly unusual in the television crime genre where good and evil, them and us, are more usually treated as explanatory categories in themselves' (Meaney, 2007, p. 10). The gender angle is another interesting feature of the production; while *The Wire* attempts to capture the male desire to provide for one's family, it also examines the strain this can cause on domestic life. Speaking of the series, Meaney highlights how *The Wire*: 'shows the desperation that drives Sobotka's nephew Nick to deal drug dealing, for example, in order to fulfil the traditional male role of providing a home for his girlfriend and child' (Meaney, 2007, p. 10).

While season one may examine mostly African Americans struggling to survive in Baltimore, season two seems to highlight the fact that it is also the Polish American and Irish Americans who are battling redundancies, unemployment and social immobility. Greek Americans are also presented in the series, in *The Wire* these characters must be ruthless and cunning in order to survive in Baltimore. Meaney also points out the fact that both genders find similar issues in Baltimore: 'While it is prominently the male role which drives the no longer working-class to crime, the women, who are strongly identified with ethnicity, are also incapable of surviving or providing for themselves in the brave new world' (Meaney, 2007, p. 10). It also seems ironic that even considering the somewhat barren landscape in Baltimore, people from foreign territories are still seeking a better life as they hope to chase after the American dream: 'the young East European women who die in the container seem to indicate that an American future is no longer available to the new arrivals' (Meaney, 2007, p. 11).

Where the two genders may differentiate in the season is probably the way in which they react to the issues present in Baltimore. One noticeable difference involves the various types of employment which is taken on by the men and women, while impoverished men are tempted by the opportunities of drug dealing, the most vulnerable women are left subject to prostitution. Outlining the grim realities for these women, Meaney states that *The Wire* introduces us to: 'an underworld of young women who cannot or will not communicate, moved around by pimps and immigration officers, with no apparent violation of their own, except their desperate desire to stay' (Meaney, 2007, p. 11). Interestingly, even women in more secure positions in *The Wire* struggle to find happiness in their personal lives as the strains of their work seem to hamper their ability to succeed. Speaking of Beatrice Russell, Meaney states that: 'her prospects for the kind of detective work which she learns to enjoy are severely restricted by the fact that she is a responsible and competent parent' (Meaney, 2007, p. 11).

Similarly, Kima Greggs finds she has no time for domestic life because of her crucial role in the police force. Considering these viewpoints, it seems that it is not only the unemployed who are struggling in Baltimore in *The Wire*, it is also those who find themselves in respectable positions, but are left weathered by either their jobs, the city or hierarchies which dominate the society. One begins to imagine that nobody is safe from the coercive effects of post-industrialisation in America.

A SPATIAL ANALYSIS OF POVERTY, CLASS AND INEQUALITY

The Wire highlights themes of poverty, class and inequality. Margaret Weir's 'From Equal Opportunity to the New Social Contract: Race and the Politics of the American "Underclass"' attempts to document the origin and history of the underclass in American society. Interestingly the topic of class has become politicised recently in America: 'During the 1980's, the language of class, in a particular form, entered mainstream political debate in the US' (Weir, 1993, p. 93). Class divide seems to be an increasingly relevant issue in America today and *The Wire* is more than capable of portraying class struggle through a wide range of characters.

While themes of poverty, class and inequality are evident in the series they are not separate issues, rather, they are linked entities requiring attention. One of the series actors, Wendell Pierce, who plays detective Bunk Moreland in the series describes how 'West Baltimore is so small and the concentration of poverty and destruction of the educational infrastructure [so intense] that it's a mass of clinical depression just now being diagnosed' (Alvarez, 2010, p. 71). The ability of the series to capture inequality is uncanny as the viewer starts to immediately question the relevancy of hierarchies, structures and bureaucracies. Those with selfish interest in the series are portrayed as: 'squeezing a buck from a system that needs people at the bottom in order to prosper' (Alvarez, 2010, p. 71).

THE UNDERCLASS AND THE POLITICS OF RACE

Weir attempts to outline the manner in which the underclass became relevant again with regard to public attention: 'The spectre of a permanent underclass caught in a web of pathology, dependent on public welfare, raised widespread alarm, focusing the public eye on the black urban poor for the first time since the 1960s' (Weir, 1993, p. 93).

One would be inclined to think that recent economic downturns and market failures have made academics and politicians alike more sensitive to class structures. Politicians may also be concerned about voting trends; they may feel that in times of economic depression, there will be a wider turn out with regard to voting. In an effort to not provoke the underclass, policy makers may attempt to create legislation that would allow for social mobility and increased redistribution of wealth.

Weir states that the underclass has created a significant impact on politics in America: 'As concern about the inner city has grown, a politics of the underclass has taken shape, complete with arguments for reorientating policy and revising national conceptions of citizenship' (Weir, 1993, p. 93). Similarly, Bilton points out the connectivity of issues in a society: 'Although social relationships of race, class and gender are often considered in isolation from each other, in practice they operate together to generate structures of power and inequality' (Bilton et al., 2002, p. 184).

Issues and features of society can also play against each other and conflict with one another: 'Similarly, gender, class and race are all potential (and sometimes competing) sources of identity and collective action' (Bilton et al., 2002, p. 184). However, one should be careful to cautiously consider the type of attention which is being garnered with regard to the underclass. It would seem that some political bodies are portraying the underclass as the most problematic people with regard to the wider American economy. Weir outlines the shifting ideologies which can occur surrounding the underclass in particular: 'In place of the ideology of equal opportunity, which provided the justification for the War on Poverty during the 1960s, is a new political formation centred around the idea of "a new social contract" which emphasizes the obligations of citizenship rather than its rights' (Weir, 1993, p. 93).

With this in mind, one could imagine some political parties in America highlighting the importance of individual choice over government intervention. In so doing, governments deflect blame for the faults of criminals or deviants away from bureaucracy and towards the underclass or whomever may be portrayed the perpetrators of the social contract. Race, as a marker of difference, can often lead to political or social cleavages: 'Race, in some circumstances, could be treated as analytical categories – status groupings competing for scarce resources and/or power' (Rex & Moore, 1967; Bilton et al., 2002, p. 184).

It is interesting to note the significant impact themes of race and class may have on American politics and national elections. Weir outlines the emergence of politics that was motivated in coordination with class relations: 'the politics of the new social contract emerged from the failure of equal opportunity, facilitating the creation of a political configuration predicated on the political isolation of African Americans and an emphasis on individual responsibility over governmental action' (Weir, 1993, p. 93). This political isolation of African Americans seems all too evident in *The Wire* as a whole African American community is depicted in the desolate projects or other run down parts of Baltimore. Some would claim that reaching out to the vulnerable in these conditions is not just for socio-economic benefit, rather, that it is in fact tied up in judicial duty. Speaking of political movements in America, Weir states that: 'The guiding notion was that of righting a wrong: the poor economic situation of blacks represented a national failure to apply the principles of justice equally to black Americans' (Weir, 1993, p. 96).

The argument of personal responsibility seems to be consistent in the ideology of right-wing parties who emphasises individual decision making over wider societal accountability. However, the argument here is that these individuals, as members of the underclass, are born into such difficult situations that they are left vulnerable to manipulation and poor living standards. *The Wire* also seems to portray some characters as suffering at the hands of institutions or authorities. It would seem evident that redistribution and reallocation of resources and time would be needed to correct economic and societal disparities in America, particularly in cities like Baltimore. However, this redistribution is simply not enough with regard to creating a more egalitarian state. As Weir has pointed out, there are racial issues embedded in American society.

Speaking of political emphasis on personal responsibility Weir states that: 'The focus on the individual and on social factors "crowded out" alternative interpretations of the economic situation of African Americans' (Weir, 1993, p. 97). It would seem that 20th century American politics relied on statistics and figures to manipulate public perception on issues like unemployment, crime and poverty. However, the agenda of some parties in government would have been to push blame onto nameless individuals as opposed to outlining wider social issues which were damaging the potential for economic success in America: 'In the individual-orientated framework which guided thinking about how to achieve equal opportunity, unemployment and underemployment in a period of prosperity were instead taken as evidence of the problems that blacks had in accepting the responsibility of work' (Weir, 1993, p. 97).

Weir's statement here is relevant as it provides an insight into the underlying agendas which may dictate political directions. In this instance with regard to employment, the issue is politicised. The various ways in which to deal and interpret the unemployment will divide voters. However, it is also a racial issue; in this case Weir is discussing the disproportionate amount of unemployment among African Americans in America in the mid to late 20th century. The political reaction of emphasising personal responsibility in the face of an issue that clearly needed wider readdressing is not unlike imperialist agenda which may set out to benefit some at the expense of others. In this case the economically advantaged can sustain growth and individual wealth, guilt free, while the vulnerable are vilified and exploited. While there may be no outwards expansion or formal colonial activity, there is a system in place which divides people. In this sense one can relate American society to imperialist culture or colonial expansionism.

When examining issues such as race, it would be important to consider how exactly it is that ethnicity affects characters in relations to other problems in *The Wire*. One could in fact claim that racism, or at least issues of race in *The Wire* are not presented in a manner that we may expect. In fact, there are not a whole lot of examples of blunt or immediate racism; rather one is exposed to a world where structures, institutes and authorities are subconsciously programmed to discriminate, whether it may be on the grounds of race, class or background. It could be argued that this type of discrimination if more sinister and challenging of expansion of human rights than blunt displays of hatred towards minorities or those of other ethnicities.

SUBTLE RACISM, INCLUSION AND EXCLUSION

As stated in previous articles, the housing projects are a good example of institutionalised discrimination or racism. There is clearly no concern shown for the residents in these apartment blocks. Political figures ignore them, policemen try their best to avoid these areas and even the drug peddlers seek only exploit the dependent drug addicts in these areas. This depiction represents a problematic feature of class divisions in a city like Baltimore. While residents in the projects may be highlighted as problematical by officials in Baltimore, one could claim that these people should actually be receiving most aid in order to restore a sense of economic and social balance to the city. It seems ironic that there are efforts made by characters in *The Wire* to actually restrict the population of the projects to their living quarters when in fact they should be working to serve the most vulnerable.

In order to understand institutionalised and modernised racism, the dissertation will examine David Goldberg's *Racist Culture: Philosophy and the Politics of Meaning*. In his book, Goldberg attempts to bring understanding to terms like race and ethnicity and their effects in contemporary society. Interestingly, Goldberg also attempts to document the changing nature of racism in line with sociological developments and historical events. The relevancy this book brings to the dissertation is that it opens up possibilities to highlight discrimination in other forms aside from direct and blunt racism which isn't as heavily featured in *The Wire*. With regard to this chapter the themes of class, poverty and inequality are closely linked to issues of race in the series.

Goldberg attempts to highlight racism as it has developed alongside modernity: 'racial definition and its attendant forms of racist articulation emerge only with the institution of modernity, and they transform in relation to the principal formative developments in modernity's self-understanding and expression' (Goldberg, 1996, p. 1). With this statement in mind, one would be inclined to believe that racism does not exist in a static form; it is changing in its nature and can take form in developments, institutes or even power structures. Continuing the idea that racism has changed forms throughout time, Goldberg claims that: 'Racial thinking and racist articulation have become increasingly normalized and naturalized throughout modernity, but in ways *not simply* determined (as dependent variables) by social conditions at specified times' (Goldberg, 1996, p. 1).

The type of racism which Goldberg is highlighting is the nature of discrimination which the book is interested and also the type of inequality which is depicted in *The Wire*. More primitive forms of racism may be seen as individual aggression towards other ethnicities, the segregation of communities or different treatment of people in a society depending on their race. However, what Goldberg is stating and what the chapter is hoping to highlight is the idea that racism does not need to be carried out by individuals or people, it can be found in systems or structures which operate in societies. The troubling issue with this embedded type of racism is that it is hard to prove and also harder to combat: 'The problem of institutional racism is difficult to tackle in organisations that ignore the evidence of tacit discrimination within their own structures' (Bilton et al., 2002, p. 174).

Goldberg identifies Liberalism as a medium through which racist agendas may emerge. Speaking of Liberalism, Goldberg states that: 'As modernity's definitive doctrine of self and society, of morality and politics, liberalism serves to legitimate ideologically and rationalize politico-economically prevailing sets of racialized conditions and racist exclusions' (Goldberg, 1996, p. 1). Goldberg further documents the origins of race and modernity and implies that since race has been understood and perceived by people, that people have used it as a marker and indicator of difference and as an excuse to discriminate: 'Race undertakes at once to furnish specific identity to otherwise abstract and alienated subjectivities' (Goldberg, 1996, p. 2).

By linking the end of formal imperial history with the beginning of new urban developments, Goldberg sees a definitive link between the two eras: 'It seems uncontroversial to claim that the roots of the racialized post-modern city can be traced to the end of the colonial era' (Goldberg, 1996, p. 187). The cities of many European or American countries often represented bureaucratic and organisational centres which were linked to a centralised imperial domain. The interest in this area of the book is how these cities function as racially diverse centres in post-colonial countries such as America. Speaking of the origin of the racialised post-modern city, Goldberg states that this marked a societal need to: 'confront directly the "problem of the racially marginalized," of (re)producing racial marginalization in its own spaces' (Goldberg, 1996, p. 187).

Goldberg continues by documenting the sometimes-changing roles of the city space. He states that at times people have been unknowingly segregated into particular zones or spaces in the city and been kept divided by parks, amenities or even modes of transport. He identifies the city as an area where racial and economic differences can be exaggerated: 'Differences within urban structure – whether economic, political, cultural or geographic – are in many ways magnified by and multiply the social hierarchies of power in and between cities, between town and country' (Goldberg, 1996, p. 185). Again, Goldberg identifies modern agents such as capitalism as facilitating the racialisation of the city as space: 'The main modes of social exclusion and segregation throughout maturing capitalism and modernity have been effected in terms of racialized discourse, with its classifactory systems, its orders and values, and its ways of "seeing" particular bodies in their natural and social relations' (Goldberg, 1996, p. 187). While one may hope to believe that the city would prove to be an area of successful racial mixing it often proves to be hampered with issues anchored around ethnic or class division. Goldberg claims that: 'the doctrine of segregation was elaborated largely with the twentieth-century urbanization of racial Others' (Goldberg, 1996, p. 187).

As the city operates acts as a divider, a separator of class and race, racism is adopted by the institutions that organise apartheid like control of the urban districts. The struggle to control the modern city originates in the mid-20th century. With wealth disparities constantly changing in America, people either fled to, or from urban centres, however in this process: 'Exclusion and exclusivity were internalized within the structures of the city planning throughout the expanding (cos)metropolises of the emergent "West"' (Goldberg, 1996, p. 188). This was also the beginning of the trend of white middle class families opting to move out from the city which was becoming dangerous in their eyes, to the more elusive suburbs. However, it is evident that decades later the pattern reversed, gentrification began to occur and young white professionals took up homes in inner city areas, as Harvey has highlighted in his study *Spaces of*

Hope. In the case of Baltimore, as presented in *The Wire* one can see how the city has been divided up into different socio-economic quarters. Regarding the housing projects, Goldberg's claims seem to be more than accurate: 'the racial poor were simultaneously rendered peripheral in terms of urban location and marginalized in terms of power' (Goldberg, 1996, p. 188). Similarly, in the series there are also efforts to gentrify some parts of the city. However, due to the crippling effects of poverty in neighbouring areas, much of the attempts to develop are stifled by an underlying disparity among the different classes in the city.

'OTHERNESS' AND THE CHANGING NATURE OF RACISM

While the essay has created links between the underclass as depicted in Baltimore in *The Wire* and a group of people who have been subject to colonisation, it could also be claimed that one is in fact viewing a representation of the 'Other' when they watch *The Wire*. It seems that this 'Other' represents something radically different from traditional American life. In the case of *The Wire* it would seem that an attempt is made to challenge the viewer's idea on what the 'Other' really is and what this culture represents in relation to homogenous identity. Spencer poses the question, 'What is the Other?', he claims that: 'the existence of a group of people – a "them" rather than an "us" – can be seen to be important in human society as it affirms qualities and characteristics that a group sees as normal as the rules by which they live' (Spencer, 2006, p. 8). While the concept of 'Otherness' here originally derives from Edward Said's *Orientalism* the book is interested in highlighting 'Otherness' within wider culture, as opposed to difference created between two distinct areas.

Spencer's statement here inspires a number of different questions which should be considered regarding both meanings of race and difference. A theme which is also evident in *The Wire* is the idea of interconnectedness among the different classes or races in Baltimore. An idea is expressed that the dominant homogenising culture actually needs an 'Other' to contrast against. One could consider scenes in *The Wire* where characters such as Omar Little attack the legal system and authorities for 'needing' crime for the sake of their own employment. Spencer continues highlighting affirmations in identity and how this can lead to the development of an 'Other' or an outsider: 'In the most obvious cases these affirmations gain official and legal status and those breaking the rules are portrayed as "criminals," "insane," "deviants" or "antisocial"' (Spencer, 2006, p. 8).

One could claim that in fact 'Otherness' is more aligned with socio-economic positioning in a society. However, with this in mind it is also evident that 'Otherness' can also be qualified purely on ethnic difference. While the 'Other' can represent a completely external body or group of people, evidently the 'Other' can appear within the 'Occident'. Spencer highlights the definitive links between the underclass and the 'Other': 'They may live and work on the fringes of "respectable" society, and make up a marginalised group or underclass, but they are also frequently associated with racial or ethnic groups' (Spencer, 2006, p. 8). Evidently, the media also plays a large role in shaping perception regarding the underclass or the 'Other'. *The Wire* may in fact be inspiring people to redefine what 'Otherness' means and how issues like poverty, class and equality should be understood. Sometimes the homogenising culture finds its definitions by pitting its codes against the 'Other': 'some of the complex negotiated views that for our self-identity derive from the media's images of "others" as well as images of ourselves as we would aspire to be' (Spencer, 2006, p. 9).

One could claim that negative depictions of the 'Other' or 'Otherness' harm potential for equality within a country or city. Similarly, social mobility may be decreased in the homogenising culture becomes too rigid in what it allows as 'acceptable'. Minority groups trying to establish a sense of financial and social stability in a city such as Baltimore may find it difficult to develop under the pressure of negative images of the 'Other' which are presented in the media. Individuals and societies often attempt to define themselves in relation to others: 'we are constantly monitoring our environment and establishing reference points, alignments with shared values and beliefs and contradistinction with others with whom we have less in common – or who are portrayed as enemies of our society' (Spencer, 2006, p. 9). However, one should be careful not to imagine identities, notions of difference and 'otherness' as static or consistent. It seems that throughout history the nature relations between different groups of people has changed drastically: 'These relations of inclusion and exclusion are far from simple and tend to be organised around different identities at different times in our history' (Spencer, 2006, p. 9).

In a chapter titled 'The Racisms of Globalization', Stephen Castles outlines the changing nature of ethnic divides in contemporary multicultural societies. He outlines how: 'the countless expressions of everyday racism, which reduce the life-chances of ethnic minorities in many countries' (Castles, 2000, p. 163). The emphasis here is that minorities not only suffer the personal effects of racism, rather, that their own socio-economic well-being is affected. The book will attempt to continue to highlight how globalisation can affect race and ethnicity. Again, considering the focus of this chapter, particular attention will be given to topics such as class, race and equality. Similar to other theorists such as Goldberg or Spencer, Castles is aware of the ever changing nature of racism: 'There are new types of discrimination and exclusion, as well as new ideologies to justify them' (Castles, 2000, p. 164). This may be seen as an issue for those who seek to promote equality and end discrimination. The reason for this being that it becomes harder to pinpoint incidences of racism and exclusion. As stated earlier in the chapter, the racism is internalised into the structures of the city and operates in a persistent but discreet ways.

It is interesting to consider who is responsible for exclusion within a society. Evidently, it seems that it is not individuals that do the exclusion themselves. However, people may act as agents who allow racism to permeate through institutes or structures in a setting. It also seems apparent that globalisation is changing the nature and the effects of racism. Racial issues in *The Wire* seem to be influenced by wider trends regarding globalisation. Castles also sees modernity as an agent or racism: 'Racism – as an ideology which justified European domination – has always been part of modernity' (Castles, 2000, p. 164). The inequality which is evident in *The Wire* presents itself in specific incidents; however, it also seems to manifest itself from within the operating structures of the city. There is a scene is season one, episode two in which McNulty and Bunk are interrogating Dee' Angelo about a killing in the city. The policemen tease the suspect and claim that he could kill anybody in the projects without raising attention. However when a citizen, especially a 'working man' gets killed there is a different reaction from the authorities and media. Even considering the somewhat exaggerated depiction in this scene, it would seem to be an accurate representation of Baltimore where citizens are ranked in accordance to their perceived importance to society. The level of inequality represented here is not just economic; rather it involves different valuations of human life.

Detailing the intertwined nature of class and race, Castles states that: 'racism should not be seen as a working-class phenomenon, but rather as one product of the current decline of the working-class culture and organization' (Castles, 2000, p. 183). The statement here is a good example of how defining issues like class and race in an isolated sense may make the fields seem distant or unrelated, however a very close analysis of the two issues will lead one to believe that the themes can be better understood alongside one another. Castles' claim regarding working-class racism would seem to be true with regard to discrimination in *The Wire*. Characters may voice out against other ethnicities not over biological difference or a perceived racial superiority, however frustrations may be vented as a result of a lack of opportunities.

One should be careful not to presume that racism stops at the working class, as Castles highlights: 'the absence of overt racism should not lead us to think that racism has declined in the middle or upper classes' (Castles, 2000, p. 183). Castles identifies racism as part of the crisis of modernity. The sometimes corrosive effects of modernity can lead to issues remolding and re-emerging with a new momentum. Speaking of the crisis of modernity, Castles states that: 'Racism should be seen not as a result of the crisis, but rather as one form of expression of the crisis' (Castles, 2000, p. 183). As *The Wire* highlights, racism can permeate its way up to the elite members of society and is not exclusive to any particular class.

With regard to inequality, the state and politics it sometimes seems that it is hard to know who to blame for exclusion or discrimination in a society. Documenting the racialisation of politics, Castles claims that neo-conservatism has influenced the spreading of modern racism: 'The attack on the welfare state helps to create the social conditions for racism, while the ideology of neo-conservatism provides a fertile climate for blaming "deviant" minorities for social problems' (Castles, 2000, p. 184). *The Wire* was released between 2002 and 2008 in the midst of neo-liberalism and incorporates Castles' critique of racist politics. Inequality and poverty fails to be addressed as a racist culture becomes rampant in society: 'Those groups which are disadvantaged or excluded must therefore be the victims of their own "inferior cultures" and of their refusal to adapt to the superior majority culture' (Castles, 2000, p. 184).

One could claim that racism has become globalised through the processes of modernity. Castles draws attention to a fundamental divide in new global order, he states that: 'Individuals or groups who possess the characteristics necessary to fit into the global markets, whether for labour, capital or cultural goods, are included into the global order as citizens, with civil, political and social rights' (Castles, 2000, p. 124). This insight sheds much light on socio-economic standing and class relations in both America and in *The Wire*. There is a particular type of citizen that is sought-after by the global order. The desired citizen will fulfil the needs of the economy; in this framework the people are expendable to the global order. Castles highlights the exclusion and discrimination which occurs under this framework: 'Individuals and groups who do not fit are excluded, and may be denied even the most basic rights, such as the right to work and the right to food security' (Castles, 2000, p. 124).

Surely such a reactive and regressive framework will stifle opportunities, equality and social mobility in any setting. This type of framework is not only limiting to individuals in a society it also stops any kind of accountability as: 'This process which includes some and excludes others appears to be the result of anonymous market forces' (Castles, 2000, p. 125). Anonymous market forces do not need to answer to citizens who have been excluded from a society for not meeting the needs of the global order and this leads to a regressive society. Under this framework of exclusion, consequences do not stop at state, or national level. A negative pattern of exclusion may become globalised as: 'No individual, institution or state therefore seems to bear the responsibility for developments which may impoverish millions' (Castles, 2000, p. 125). The harsh realities presented in *The Wire* would illustrate the contemporary reality that those ready to meet the needs of the globalised economy will succeed, those less well equipped, socially, culturally or politically, are destined to fail.

CONCLUSION

This chapter has examined a number of different issues and addressed the content of *The Wire*. The first section addressed the importance of race and representation. By examining these topics the chapter highlighted how TV and media representations can influence public opinion and impact people's identity. The concept of race and representation was also examined alongside a number of different writers and theorists. The politics of representation was also highlighted as an important factor when examining a TV series in a critical manner. The second section examined the economic disparities presented in the series and attempted to outline how social mobility was hampered by a dysfunctional economy. Similarly, the idea of internal colonialism and the underclass was discussed with the alongside the content of *The Wire* in order to provide an understanding of urban developments. Meaney's work, which documented issues regarding class and gender, set out to highlight how economic difficulties manifested in the characters personal lives and hampered their standard of living.

The study later returns to the idea of race but is particularly focused on ideas regarding racism and discrimination. The final section identified how class was not just a social phenomenon, but also a political one which could potentially swing voting trends. The political divide of personal responsibility and governmental provision was discussed in light of themes like class and race. Goldberg's theories on race and racism helped structure the arguments outlining the type of discrimination which was evident in *The Wire*.

Spencer's idea's on 'Otherness' in the urban centre helped contextualise issues presented in *The Wire*. The final section highlighted how different classes may either experience or exhibit racism in varying ways. Similar to other theorists used in the book, Castles stated that race was a very political issue which can become exaggerated in the media. *The Wire* proved to be a very rich study regarding how much academic analysis could be used to complement its content. The series therefore served as an interesting text which could accurately represent a spatial analysis of crime in contemporary society.

REFERENCES

Alvarez, R. (2010). *The Wire: Truth be told*. London: Canongate.

Bilton, T. et al. (2002). *Introductory sociology*. London: Palgrave Macmillan.

Blauner, R. (1969). Internal colonialism and ghetto revolt. *Social Problems*, *16*(4), 393–408.

Castles, S. (2000). *Ethnicity and globalization: From migrant worker to transnational citizen*. London: Sage.

Fraley, T. (2011). A man's gotta have a code: Identity, racial codes and HBO's The Wire. *Darkmatter*, May 29. Retrieved from darkmatter101.org. Accessed on August 6, 2012.

Franklin, J. (2011). Common ground: The political economy of The Wire. *Darkmatter*, May 29. Retrieved from darkmatter101.org. Accessed on August 6, 2012.

Fuggle, S. (2011). Short circuiting the power grid: The Wire as critique of institutional power. *Darkmatter*, May 29. Retrieved from darkmatter101.org. Accessed on August 6, 2012.

Goldberg, D. T. (1996). *Racist culture: Philosophy and the politics of meaning*. Oxford: Blackwell Publishers.

Kelly, L. W. (2011). Casting The Wire: Complicating notions of performance, authenticity, and 'Otherness'. *Darkmatter*, May 29. Retrieved from darkmatter101.org. Accessed on August 6, 2012.

McNeil, D. (29 May, 2011). *White Negros and The Wire*. Darkmatter. Retrieved from darkmatter101.org. Accessed on August 6, 2012.

Rex, J., & Moore, R. (1967). *Race, community and conflict*. London: Oxford University Press.

Sharma, A. (2011). Editorial: 'All the pieces matter' – Introductory notes on The Wire. *Darkmatter*, April 29. Retrieved from darkmatter101.org. Accessed on August 6, 2012.

Spencer, S. (2006). *Race and ethnicity: Culture, identity and representation*. London: Routledge.

Tosacno, A., & Kinkle, J. (2009). *Baltimore as world and representation: Cognitive mapping and capitalism in The Wire*. Dossier, April 8. Retrieved from dossierjournal.com. Accessed on August 6, 2012.

Weir, M. (1993). From equal opportunity to 'The New Social Contract': Race and the politics of the American 'Underclass'. In M. Cross & M. Keith (Eds.), *Racism, the city and the state*. London: Routledge.

TOWARD A SPATIAL ANALYSIS OF METHAMPHETAMINE IN NORTH AMERICA

Kevin Bucciero

ABSTRACT

This chapter will develop a comprehensive review of methamphetamine usage with a concentration on North America. To develop a wider understanding of the use of this drug, this chapter will examine the history and spatial component of methamphetamine. In order to achieve this, the research will review previous literature, which includes the pattern of its original manufacture in the United States to international locations such as Mexico. The research data will include two interviews with individuals involved in the underworld of meth use in Los Angeles. The chapter will then present outcomes based on the participant survey and secondary sources.

Keywords: Methamphetamine; border; addiction; spatial; urban; drug-use

INTRODUCTION

Amphetamine-type stimulants, which include methamphetamine and amphetamine, were developed as artificial replacements to ephedra. Ephedra is a botanic extract of *Ephedra sinica* and has been used in traditional Chinese medicine as Ma Huang for over 5,000 years (Vearrier, 2012). The search for a

Environmental Criminology: Spatial Analysis and Regional Issues
Advances in Sustainability and Environmental Justice, Volume 20, 135–140

ISSN: 2051-5030/doi:10.1108/S2051-503020170000020008

synthetic ephedrine substitute resulted in the development of amphetamine-type stimulants, produced via modification of the ephedrine skeleton with the pharmaceutical goals of CNS stimulation. Japanese chemist Akira Ogata first synthesized methamphetamine in 1919 using ephedrine as a precursor (Vearrier, 2012). Methamphetamine, being street named, meth, crystal, or ice, is a central nervous system narcotic that can be injected, smoked, snorted, or ingested orally. Long-term use at high levels can result in addiction (Anglin, Burke, & Perrochet, 2000).

Methamphetamine was used by the military. In World War II, methamphetamine was available to military personnel as Pervitin in Germany and Philopon in Japan (Vearrier, 2012). Methamphetamine in Germany and Japan and amphetamine use in the United States were used to increase alertness, reduce fatigue, and suppress the appetite of soldiers. Civilian use of methamphetamine was commonly used as a treatment for depression and obesity. The stimulate reached its peak at 31 million prescriptions in the United States in 1967 (Anglin et al., 2000). Effects of using ice can include feeling of euphoria, hallucinations, sleeplessness, cramps, trembling, paranoia, nervousness, stroke, heart attack, and structural changes to the brain (Anglin et al., 2000). Until the late 1980s, illicit use and production of ice was mostly isolated to California, but the drug user population has widened in North American with time. Notably via American media, an amplified use occurring in the Midwest of the United States. Over 4 million Americans are estimated to have tried ice at some time in their lives (Anglin et al., 2000).

LITERATURE REVIEW

From Made in America to Hecho en Sinaloa

Al Gilbreath's PhD thesis begins with statement that most of the major drugs of abuse in the United States have a relatively uniform distribution. Their use may concentrate in cities, for example, but that overall pattern tends to repeat itself in every region of the county. This is not true of the stimulant methamphetamine, which today shows a decidedly uneven distribution (Gilbreath, 2012). Confusing the matter more is the fact that because it is a synthetic drug, it is supposedly possible to make methamphetamine anywhere. But it is not made everywhere. In fact, for much of its history, the drug has been concentrated in the American West (Gilbreath, 2012).

Meth laboratories can be built in homes, tents, garages, and car trunks. "One pot methods" only need a soda bottle. The potential sites for manufacture are extensive. Conversely, methamphetamine production is regional. In 2003, only three methamphetamine labs were seized in all of New England. New York and New Jersey had only 10 labs combined. For the sake of comparison, 82 lab seizures occurred that year in a single Missouri county – Jefferson. The entire state of Missouri had over a thousand lab seizures in 2003 (Gilbreath, 2012). Over the progression of the last two decades, ice usage has begun to travel from West to East across the North America. Nevertheless, currently the drug is randomly scattered around North America, mostly leaving the east coast untouched.

Mexican meth started in the late 1980s when U.S. Congress passed the chemical diversion trafficking act. Its main goal was to stop U.S. companies from exporting chemicals needed to make ice and cocaine. In particular, the chemicals ephedrine and pseudoephedrine. Mexican drug cartels moved in when domestic production slowed. The geography of methamphetamine that was recognized in the 1980s has now been redirected in the location of the Mexican drug syndicates today. The organizations that control the border are involved with methamphetamine production and distribution (Gilbreath, 2012).

WEST COAST BOOMS AND EAST COAST BUSTS

Gilbreath asserts, in his introduction to his 2014 article that 77.8% of methamphetamine labs seized in the country between 2000 and 2003 were found in the West Central, Pacific, and Southwest (Gilbreath, 2014). His article seeks to explain the uneven western circulation of methamphetamine manufacture and distribution. The author finds that the drug's western origin dates back to the regulation of a particular precursor ingredient, phenyl-2-propanone (P2P), in 1981. In response to the law changing, producers on the West Coast developed a new process of making the drug, which resulted successfully in both production and sales in that region. East coast meth, which was associated with traditional organized crime dependent on trafficking the "P2P" from abroad. This method failed from police interception (Gilbreath, 2014).

> The first major dealers of ice in the 1970s and 1980s were the outlaw motorcycle gangs such as the Hell's Angles. They can be considered efficient in drug trafficking for the reason of mobility. It's simple for the gangs to get on their motorcycles and drive out of state at any time, while the local police can't go outside their precincts. An anonymous member of bike gang explained the importance of the drug: Everything in the whole club revolves around crank. You can't ride a $10,000 motorcycle, have a big gun collection, and take care of three 19-year-old ladies working in no body shop. (Gilbreath, 2014)

MAPPING THE SPREAD OF METHAMPHETAMINE: CALIFORNIA 1995–2008

The intention of this chapter is to conclude that a drug epidemic has occurred during the selected time period in California. This study researches the amplified incidence of meth abuse that hospitals have reported. Patients intoxicated with methamphetamine have increased 13 times since 1983–2008 (Gruenewald et al., 2013). This research, which was published with the *American Public Health Journal*, used the methodology of geocoding hospital discharge data to residential zip codes from 1995 to 2008. The discharges were related to population and environmental factors using a statistical model of Bayesian Poisson. This enabled assessment of geographic areas where the abusers lived. The results showed three increases phases, then a disruption in the time period when laws were enacted to limit meth production precursor chemicals available to the public. Outcomes also displayed meth abuse was greatest among the White and Hispanic low income population. Other notable characteristics of patients included living in areas near a highway system, having a small household size, and living outside dense urban areas. This article concludes that during the analyzed time period the large increase in meth abuse hospitalization in California exhibited features of a drug epidemic.

SPATIAL MODELS: THE GROWTH AND SPREAD OF ICE

The spatial models research also by Gruenewald studies the severe and increasing public health issues related to methamphetamine abuse in the United States. It combines economic and scientific methods to explain the movement of ice abuse, treating the stimulant use as a continuing and worsening disease that moves through society with the assistance of the black markets. Gruenewald's models suggest that methamphetamine might display classic disease characteristics such as spatial clustering and related growth (Gruenewald, 2014). These models were verified using historical records on methamphetamine-related arrests and hospital discharges in California between 1980 and 2006. The information proposes that both problem signs increased significantly during this period excluding for short-term source reductions after the enactment of federal bans on the precursor substances used to produce methamphetamine. The spatial movement of methamphetamine abuse was studied using Bayesian space-time models of arrest counts in 330 California cities (Gruenewald, 2014). The results showed that cities ranged substantially in both their arrests and their increase rates over studied period.

These changes were also seen among nearby cities, as forecasted by the public health approach by which a methamphetamine "infection" spreads from person to person. These results show that methamphetamine growth was highest in rural northern and southern California between 1980 and 1989, then moved to the central valley areas during the early 1990s before moving more into the urban areas after 1997 (Gruenewald, 2014).

DATA

I contacted the LAPD Narcotics Division by telephone and was connected with a Detective in The Gang and Narcotics Division, NAU 2 and Medical

Marijuana Business. The following is answers to a Five-Question Survey by the Detective.

Survey 1: (Answers in bold)

1. In your jurisdiction, where is the majority of the methamphetamine produced?

Mexico

2. In your jurisdiction, what percent of methamphetamine is produced domestically?

Potentially less than 20% is processed in the outlying areas of LA County, but up to half is "cleaned up" locally. We rarely see meth labs in town anymore. The bigger problem now is THC extraction labs which use butane and cause houses to go "boom." Meth often arrives "dirty" with impurities or too much "cut" present, from Mexico. It will have a brownish peppery appearance. Acetone or alcohol is most commonly used to remove the impurities and generate new crystal formations.

3. In your jurisdiction where was the majority of methamphetamine produced 10 years ago?

About 40/20 outlying areas and in residential areas. Rest was initial startup of Mexican labs.

4. In your jurisdiction what percent of methamphetamine was produced domestically 10 years ago?

Approximately 60%

5. In your jurisdiction what significant changes in methamphetamine are being seen? "Geographic trends/movements, physical form, etc."

More of it has come in thanks to NAFTA, which has allowed more commercial trucks from Mexico to cross the border, and the availability of precursor chemicals in Mexico. Lower level traffickers use hidden compartments ("Traps") to hide in their vehicle and transport it across the border. The form/appearance hasn't changed a whole lot but the purity has dropped a bit out of Mexico and the price is down locally. The east coast it goes for 4–5 times what it goes for in So. CAL, which is $2,300. That is a price drop of @ 21% since 2015. Once here, the meth is transported by truck or in luggage to the east coast, Chicago, Detroit, etc.

Survey 2:
While walking in downtown Los Angeles, solicitors often whisper a street drug name to offer a sale. I chose to interview one of the drug dealers on October 12, 2016. "E" answered four questions from my survey.

1. Where does your "crystal" supply come from?

The street, its cooked

2. What drug has the most demand in L.A.?

(1) Cocaine (2) Meth (3) Heroin

3. How much is your monthly income from meth?

120 a month but other dealers can make 2,000–3,000

4. What changes have you seen in the drug business in the last 10 years?

More users (Survey Ends)

CONCLUSION

This chapter's inclusive review of methamphetamine usage concentrated is on North America. In direction to develop a wider understanding of the use of this drug, the chapter regionally investigated Los Angeles and explored the history of methamphetamine. This chapter's goal was to achieve an understanding of meth with previous literature review, extending from its initial production in the United States to abroad. This study data included dialogs with people connected to the world of drug use in Los Angeles. One outlaw individual and one in law enforcement. A conclusion that can be drawn from this chapter is that meth's future comes from United States' southern border. The North American Free Trade Agreement's goal is to make trading between the economies easier. NAFTA has also given the Mexican drug cartels an easier route to traffic methamphetamine. The current cost of ice is significantly higher on the east coast making the west coast and mid-west a presumably larger market.

REFERENCES

Anglin, M. D., Burke, C., & Perrochet, B. (2000). History of the methamphetamine problem. *Journal of Psychoactive Drugs*. doi:10.1080/02791072.2000.10400221

Gilbreath, A. L. (2012). *From made in America to Hecho en Sinaloa: A historical geography of American methamphetamine networks*. Los Angeles: University of Kansas, Department of Geography. doi:462.34763487.33

Gilbreath, A. L. (2014). *West coast booms and east coast busts: Explaining the roots of methamphetamine's peculiar geography historical geography*. Retrieved from ejournals.unm.edu. doi:10.23123/873753.734657

Gruenewald, P. J. (2014). Spatial models of the growth and spread of methamphetamine abuse in California. In *Crime, HIV & health: Intersections of criminal justice and public health concerns*. Los Angeles: Springer. doi:09.1204/LHSE.2014.73344

Gruenewald, P. J., William, R. P., Lillian, G., Remer, L. A., Waller, L. Z., & Gorman, D. M. (2013). Mapping the spread of methamphetamine abuse in California from 1995 to 2008. *American Journal of Public Health*, *103*(7), 1262–1270. doi:10.2105/AJPH.2012.300779

Vearrier, D. (2012). *Methamphetamine: History, pathophysiology, adverse health effects, current trends, and hazards associated with the clandestine manufacture of methamphetaminex*.

INDEX

Printed and bound by CPI Group (UK) Ltd, Croydon, CR0 4YY

13/09/2026

14961058-0007